Science, Philosop
Physical Geography

This accessible and engaging text explores the relationship between philosophy, science and physical geography. It addresses an imbalance that exists in opinion, teaching and to a lesser extent research, between a philosophically enriched human geography and a perceived philosophically empty physical geography.

The text challenges the myth that there is a single self-evident scientific method that can, and is, applied in a straightforward manner by physical geographers. It demonstrates the variety of alternative philosophical perspectives and emphasises the difference that the real world geographical context and the geographer make to the study of environmental phenomena. This includes a consideration of the dynamic relationship between human and physical geography. Finally, the text demonstrates the relevance of philosophy for both an understanding of published material and for the design and implementation of studies in physical geography.

This edition has been fully updated with two new chapters on field studies and modelling, as well as greater discussion of ethical issues and forms of explanation. The book explores key themes such as reconstructing environmental change, species interactions and fluvial geomorphology, and is complemented throughout with case studies to illustrate concepts.

Rob Inkpen is a Principal Lecturer in Geography at the University of Portsmouth. He has research interests in stone and rock weathering and the philosophy of physical geography.

Graham Wilson is Senior Lecturer in Physical Geography at the University of Portsmouth. He has research interests in Quaternary climate and sea-level change.

Science, Philosophy and Physical Geography

Second edition

Rob Inkpen and Graham Wilson

Routledge
Taylor & Francis Group

LONDON AND NEW YORK

First edition published 2005
by Routledge

Second edition published 2013
by Routledge
2 Park Square, Milton Park, Abingdon, Oxon OX14 4RN

Simultaneously published in the USA and Canada
by Routledge
711 Third Avenue, New York, NY 10017

Routledge is an imprint of the Taylor & Francis Group, an informa business

British Library Cataloguing in Publication Data
A catalogue record for this book is available from the British Library

Library of Congress Cataloging in Publication Data
Inkpen, Robert, 1964–
Science, philosophy and physical geography / Rob Inkpen and Graham Wilson. -- Second edition.
      pages cm
   1. Physical geography—Philosophy.   I. Title.
   GB21.I55 2013
   910′.0201—dc23

                                                                    2012049007

ISBN: 978–0–415–67965–7 (hbk)
ISBN: 978–0–415–67966–4 (pbk)
ISBN: 978–0–203–80634–0 (ebk)

Typeset in Times New Roman
by RefineCatch Limited, Bungay, Suffolk

MIX
Paper from
responsible sources
FSC
www.fsc.org   FSC® C013604

Printed and bound by CPI Group (UK) Ltd, Croydon, CR0 4YY

Contents

List of figures and tables

Figures

Tables

Preface to the first edition

A book about philosophy in physical geography has a hard audience to impress. Experts will have their own views on the key figures in the development of ideas, whilst novices will not be clear why this sort of topic is of any importance for studying the physical environment. It may all be very interesting, but how does it help me study seismically triggered landslides? I would answer that identification, classification and analysis of such phenomena is an intensely philosophical practice, it is just that much of the philosophy is invisible to the practitioner. This book will, I hope, make some of this underlying philosophical basis visible.

Although the book is aimed at second- and third-year undergraduates, I have tried to ensure that there is sufficient depth of material to make the book of use to postgraduates and interested researchers. This book is not intended to be a complete review of the existing literature of philosophy or indeed of philosophy in physical geography. Such an undertaking would be huge and require a great deal more time and space than I have available. Inevitably, this means that some texts are not included and some readers may find offence in this. I am sure that 'How can you talk about philosophy without mentioning such and such?' will be a common lament. Although it would have been nice to include detailed accounts of the thoughts of the 'great' men of physical geography, this was not the aim of the book. This is not to denigrate the debates that have developed in physical geography nor the influence of the main characters in these debates. Focusing on individuals and their 'pet' projects can drive any synthesis into the particular 'camps' of interested parties. I am not immune to pushing a particular view of science and practice in physical geography, but I hope that the reader can use the information provided to develop their own judgement of both my opinions and those of the individuals involved in recent debates.

I am not trying to state which of the competing philosophies is 'best' nor the 'correct' approach every researcher should undertake. You may be surprised in reading the contents page that I have only a single chapter covering all the philosophies (i.e. logical positivism, critical rationalism, critical realism and critical pragmaticism). This is a deliberate ploy. Comparing the different philosophies may be useful in identifying the limitations and potentials of each, but it can also imply that one philosophy is 'better' in all circumstances than any other philosophy. Although I have a personal view on what philosophy seems the best, it would be inappropriate of me to suggest that this is the only approach to take and that an easy translation of this philosophy is possible in all situations. This does not mean that I have avoided emphasising particular philosophical viewpoints. My hope is that individuals will use the questioning approach of the book to assess my views as well. This book does not supply any definitive answers, instead it supplies views and questions to ask of those views.

I identified key themes that need to be questioned by those practising physical geography. The key themes can be summarised by the questions:

- What is the reality that physical geographers think they study?
- What are the things physical geographers identify as their focus of study, and why those?
- What counts as a valid explanation in physical geography?
- How do physical geographers engage with reality to derive information from it?

Whatever philosophy you prefer you will still have to address these key questions. For each of these questions I have presented a chapter that shows that they cannot be answered simply. If you expect a clear and definitive answer to each question, a simple template to guide your research, then I am afraid you will be disappointed. No simple solutions are provided. Instead I try to make it clear that each question needs to be thought about. Each theme has no 'natural' and obvious solution. If you read each chapter and come to the conclusion that what you already do provides adequate answers to the problems raised, then that is fine. What I hope I have done is at least raised as issues what may not have appeared to be problematic. If you at least leave a chapter thinking how the issues raised might be applicable to what you do then I would regard that as a success. My purpose is not to alter practices to a single correct approach, but to raise awareness of problems and issues that appear a 'natural' and normal part of working practices.

Beginning this book with a long chapter on the history of the subject may seem an odd way to go about discussing these questions. Initially, I was not keen to write an historical chapter. Once I began to write it, however, I could see how the impossibility of writing 'a' history of physical geography brought to light a lot of the problems of 'naturalising' ideas and practices that contemporary physical geography also suffers from. Trying to illustrate these problems immediately with contemporary concepts may have been more difficult.

Although the title of this book is concerned with physical geography, it has not been possible to cover this vast and growing subject to anything approaching a satisfactory depth. Almost by default I have tended to fall back on examples and ideas within my own particular subdiscipline, geomorphology, and then even further back into my two areas of research, landslides and weathering. This is not to suggest that these two areas are the most important in physical geography or even within geomorphology. It is just that I happen to know the literature in these fields in greater detail than in other parts of physical geography. If you know of examples from your own specialism that are better illustrations of the points I am making then I would regard that as a success as well. It would mean that you have interpreted the ideas and applied them elsewhere in the subject.

Some readers will be surprised at the omission of certain topics that appear to be important topics in their subject area. Probabilistic explanation, for example, is not covered explicitly. In the case of this topic, Chapter 4 does provide the explanatory tools for understanding its structure. Modelling is not dealt with in detail as a form of explanation, although once again the tools for interpreting this type of explanation are provided. Similarly, where computer modelling is dealt with it is only briefly. The omission of certain topics reflects my choices as author. I could have made passing reference to probabilistic explanations at specific points in the text. This would have been a token effort. Such a token reference would have misrepresented important concepts that would require a long and complex discussion. I leave that to more appropriate subdisciplines. My concern has been to provide an appropriate framework and set of intellectual tools for criticising theory and practice in physical geography.

As you plough through the book it may be helpful to bear in mind that whatever physical geographers think of philosophy, some people will always want to question the basis of what researchers do. It is vital to get the whole thing into perspective. Philosophy should be of interest to physical geographers because undertaking investigation of the physical environment will involve philosophical decisions and debates, even if they are not recognised as such. Don't worry about the philosophy too much. Bear in mind the view of Vroomfondle in Douglas Adams's *Hitchhiker's Guide to the Galaxy* – maybe the best we can hope for in our studies is to 'Demand rigidly defined areas of doubt and uncertainty'.

Rob Inkpen
Portsmouth, August 2003

Preface to the second edition

Why do this again? This is the key question that anyone undertaking a new edition should ask . . . Has anything changed since the first edition was published to make a second edition anything more than an afterthought? The reception of the first edition of this book was quite positive but some key issues seemed to repeatedly be brought up. A book on the philosophy of physical geography was decidedly 'geomorphological' in flavour. The key examples and concepts were often interpreted and exemplified using geomorphological examples. Co-authorship of the second edition should prevent such a heavy bias to one part of physical geography. Having two authors from different parts of the subject meant that ideas could be discussed and agreement over 'good' and 'bad' bits from the first edition thrashed out to make, we hope, a better overall product. Satisfying two perspectives will, we hope, make the second edition more interesting and have greater use to all of physical geography and not just interested geomorphologists.

A second edition is also an opportunity to add to existing chapters any debate or new ideas that have cropped up since the first edition. The continued and slow integration of physical and human geography, for example, particularly through the emergence of 'citizen' or 'democratic' science has been an important research direction since the first edition. Some chapters have been expanded to make explanations clearer or to add examples from the fields of ecology and environmental reconstruction. Two new chapters have been added on 'The Field' and 'Modelling'. The first reflects an increasing interest in the arena within which physical geography is practiced and the recognition that this is not a simple or uncontested place. The discussion of the conceptual aspects of the field could have gone deeper into recent developments in human geography but we felt only an initial introduction to this topic was needed here. The modelling chapter plugs a clear gap in the first edition and one that has become increasingly obvious as modelling continues to occupy a prominent position in physical geography. Outlining model types introduces the reader to something that many others have already done. The critical second part of the chapter, on the issues with modelling, provides an account of some of the key constraints on modelling reality. Modellers may hold different opinions about the degree to which models are constrained, but recognition of the nature of this is, we believe, important in a world that increasingly uses models to represent and make decisions about an unruly reality.

Anyway, enough of the preliminaries: we hope that you read the book and find it informative and as enjoyable as a book about the philosophy of geography can be!

Rob Inkpen and Graham Wilson
Portsmouth, November 2012

Acknowledgements

Revision of a textbook is not the easiest of tasks and this work could not have been completed without the help and support of friends and colleagues at Portsmouth and elsewhere. At Routledge, we would like to thank Andrew Mould for suggesting a revision and Faye Leernik for her continued support and gentle prodding towards completion. In writing the revision we would like to thank our families for their support and understanding during the process. In particular, Graham would like to thank Lesley, Molly, Evie and Connie for their love and support and Rob would like to thank Shirley for her love, support and encouragement.

We gratefully acknowledge the permissions below:

A version of Figure 2.1 was published by V.R. Baker (1999) Geosemiosis. *Bulletin of the Geological Society of America*, **111**, 633–645.

Figure 5.2 is reproduced with from S.N. Lane and K.S Richards (1997) Linking river channel form and process: Time, space and causality revisited. *Earth Surface Processes and Landforms*, **22**, 249–260 with permission John Wiley and Sons Limited.

Figures 4.8 and 4.9 are reproduced from R. Inkpen and G. Wilson (2009) Explaining the past: Abductive and Bayesian reasoning. *The Holocene*, **19**, 329–334 with permission from SAGE Publications.

Figures 5.3 and 5.4 are reproduced from T. Buffin-Belanger, A. Roy and A. Kirkbride (2000) On large-scale flow structures in a gravel-bed river. *Geomorphology*, **32**, 417–435 with permission from Elsevier.

Figures 8.5, 8.6 and 8.7 are reproduced from K. Fryirs (2012) (Dis)connectivity in catchment sediment cascades: A fresh look at the sediment delivery problem. *Earth Surface Processes and Landforms*, DOI: 10.1002/esp.3242 with permission from John Wiley Limited.

Figure 8.9 is reproduced from J. Phillips (1995) Nonlinear dynamics and the evolution of relief. *Geomorphology*, **14**, 57–64 with permission from Elsevier.

Figure 9.1 is reproduced from I. Shennan, S. Hamilton, C. Hillier and S. Woodroffe (2005) A 16 000-year record of near-field relative sea-level changes, northwest Scotland, United Kingdom. *Quaternary International*, **133–134**, 95–106 with permission from Elsevier.

Introduction

Philosophy in physical geography is done as much by boots and compass as by mental activity. Philosophy in physical geography is an active process open to change as the subject is practiced. This book taps into philosophy in action in physical geography by exploring the links between the two and their relations to scientific thought as a whole. Recent developments in geography have highlighted the need for a source of information on the philosophical basis of geography. Texts on the philosophical content of human geography are relatively common (Gregory, 1978; Kobayashi and Mackenzie, 1989; Cloke *et al.*, 1992). Texts that serve the same purpose for physical geography are thinner on the ground. Apart from the classic Haines-Young and Petch (1986) text on physical geography, there is little accessible to undergraduates that discusses how physical geographers think about and use philosophy within their work. Some physical geographers may regard this as a good thing, automatically stating the Chorley (1978) quote about reaching for their soil auger when they hear the term philosophy (although the rest of the text beyond this infamous quotation is in favour of a more thoughtful approach to the subject). Physical geography has been and still is about 'doing' rather than theorising in the passive sense of 'armchair' geography, an attitude with which many physical geographers interpret the term. This book is not intending to dispute the active and reflective nature of physical geography, in fact this book is only possible because physical geography is a subject that views itself as first and foremost about practice. The activities of physical geographers, the practice of their subject, involves a vast array of philosophical decisions, even if most are implicit or seen as part of the tradition or training associated with a particular field. Importantly, physical geographers, as all good scientists, question what they are looking at – they do not necessarily take the world for granted as it is. They define, they classify, they probe, they question and they analyse. Why is this not seen as philosophical?

The lack of explicit recognition of the importance of philosophy in physical geography may arise from its practitioners' reflection on what their human colleagues regard as philosophical. Human geography seems to have different underlying philosophies, seemingly many more than physical geography. Human geographers seem to be forever embroiled in the latest debate about the nature of their reality, the construction of worlds, structuralist versus post-structuralist thinking as well as the postmodernists' debates on the nature of geography. By extension, the number of books addressing explicitly philosophical issues in subdisciplines of human geography is legion by comparison to physical geography. Each faction seems to have its own clique and its own literature. Physical geographers are caricatured as having only one philosophy, the scientific approach. By implication, such an outdated and simplistic philosophical outlook cannot compete with the philosophical sophistication of human geography.

Physical geographers have not necessarily done much to correct this misconception of the scientific approach. Many physical geographers have preferred to retreat into their subject matter, dealing with the detail of their dating methods or the representativeness of their sampling methods, leaving the philosophising to those who do nothing practical.

There has, however, been a set of wide-ranging philosophical debates that have and are occurring over the nature of scientific investigation. Most physical geographers have been reluctant to enter into this debate in earnest. The 'physics envy' identified by Massey (1999) implies that only the hard sciences can develop philosophies that the 'softer' environmental sciences of geology and physical geography will then dutifully pick up and operate with. Fortunately, recent debates in physical geography with participants such as Frodeman (1995, 1996), Baker (1999), Demeritt (1996, 2001), Rhoads and Thorn (1996a) and Richards (1990) have shown that physical geographers and their 'practical' approach to the study of the environment have as much to contribute to the developing philosophy of science as do quantum physicists. Given the highly detached and impractical nature of theoretical quantum physics, it could even be argued that physical geographers can contribute more to the debate because they operate within a discipline that has, as its subject matter, the physical environment at a more human scale.

Physical geographers have not been aphilosophical, but rather philosophy-shy. At the undergraduate level, the glaring exception to this has been Haines-Young and Petch's (1986) *Physical Geography: Its Nature and Methods*. Published in the mid-1980s and with an agenda that favoured critical rationalism as *the* scientific approach, this book provides a very general and readable introduction to some of the key philosophical questions that physical geographers need to address. At a more advanced level is von Englehardt and Zimmerman's (1988) *Theory of Earth Science*. This advanced text gave a very Western European view of the philosophical basis of the earth sciences, and covers much of the same ground as the current text. The detailed level of the argument, however, was more complicated than might be expected of undergraduates unfamiliar with the idea of a philosophy in physical geography. More recent publications have taken an historical view of the development of the subject (e.g. Gregory, 1985) or have been collections of papers such as Rhoads and Thorns (1996b) *The Scientific Nature of Geomorphology*. The latter publication provides an interesting and thoughtful introduction to thinking in geomorphology appropriate for the postgraduate level. It does not, however, provide a coherent and integrated text that explores the recent debates about philosophy in physical geography.

In the nearly three decades since Haines-Young and Petch published their text, the nature of philosophy in science and, in particular, in environmental science, has been debated and has changed. Critical rationalism still stands as a beacon for understanding reality, but the nature of that reality has been subject to debate, as has been how environmental scientists should or do explore and explain it. The caricature of science as a monolithic philosophy with a rigid and singular approach to its subject matter is one that no longer stands up to scrutiny. In its place is a series of philosophies that share a belief in an external reality. Where these philosophies differ is in their claims about the ability of researchers to obtain an absolute knowledge of reality. Critical rationalism, critical realism and pragmaticism have all been flagged in the scientific and geographic literature as being of relevance. In addition, the development of a more sociological view of scientific practice as illustrated, for example, by the studies of the sociology of science from the Edinburgh school, has not really been addressed within physical geography. Likewise, the potential import of ideas such as Actor Network Theory (ANT) for understanding the context of practice within physical geography is relatively underdeveloped. One of the things that has become clear is that it is the practice of a discipline that aids the development of its philosophy. Chorley may have been right to reach for his soil auger, but only because it is by the practice of a discipline that anyone can hope to understand its philosophy and, in turn, to understand how and why it changes.

Structure of the book

The chapters in this textbook are arranged to lead the reader through old and new ideas in the philosophy of physical geography in what we see as a logical manner. Chapter 1 deals with the context of the current debate by outlining an historical appraisal of physical geography. Whilst recognising that any single history of physical geography is impossible, there are still themes that can be teased out of the past that individual physical geographers would accept as extant in their subject. Such themes include the search for universal rules or laws, the emphasis on practice and the empirical, a concern with time, space and, ever problematically, with scale. Additionally, there has been an overriding concern to be seen to be scientific and a reverence for the 'hard' sciences.

Chapter 2 focuses on the nature of the reality that is studied by physical geographers. The central feature of the diverse 'scientific' philosophies covered is a belief in a common and externally existing, independent, reality that is capable of study by sensors. Arguments are developed for the basic assumption that lies at the heart of physical geography. The various ontology and epistemology of the philosophies used to study and explain reality are also outlined. Central to a physical science view of reality is the idea that reality is not understood as a separate entity waiting for the discovery of its absolute content. Rather, reality is understood through an ongoing dialogue with the researcher. Reality is understood only through this dialogue with all its potential misinterpretations, and is always susceptible to renegotiation.

Chapter 3 looks at the important components of any study, the entities thought to make up the external reality. How these entities are identified, classified and then studied is a major area of debate between philosophers of science. The existence of natural kinds, or the human construction of entities, and the implications of both of these for the study of reality, are of central importance to a field science such as physical geography, where the boundary between entities and kinds can often become blurred and dynamic.

Chapter 4 builds on the previous two chapters by looking at the different forms of explanation that are acceptable within science to understand how the entities identified behave. The different scientific modes of explanation each assume different things and each will accept different forms of evidence as valid for understanding. A brief critical outline of each is provided.

Chapter 5 looks at how reality is studied, how the dialogue with reality is actually carried out by physical geographers. The key relationship between the entity being measured, its context, and the measuring instrument is outlined. From this discussion, the idea of an inseparable measurement system and a model-theoretic view of science is discussed.

Chapter 6 explores the arena of practice in physical geography: the field. Physical geography often describes itself as a field science and the nature of this definition is explored and discussed. The 'field' means many things and take on many guises in physical geography. Rather than a single, separate and identifiable thing, the field is something that the researcher creates and explores on their terms, and is sometimes a wholly constructed reality studied in laboratory and experimental research. Undertaking research does not necessarily just involve one type of 'field'. Different research creations can be linked together by the researcher as they move from one type of field to another in their search for explanation.

Chapter 7 begins the exploration of how the general, philosophical concepts developed in the preceding chapters have been applied within physical geography. This chapter discusses systems analysis in physical geography. Specifically, it deals with how systems analysis has been operated within physical geography, how this operation has required detailed philosophical consideration (although it may not have been recognised as such) and where the problems with such an approach lie.

Chapter 8 explores the debates concerning frameworks for explaining change and complexity in physical geography. Recent conceptual developments, such as chaos theory and complexity, have seemed to offer much to our understanding of reality. The use of these new concepts has, however, involved both a debate over the problems of an old concept, equilibrium, as well as the reinterpretation of new concepts, chaos and complexity, within the discipline of physical geography. The current debate clearly illustrates that new and old ideas tend to mingle rather than compete. The nature of this process varies, depending on the subdisciplines involved: there are no fixed definitions of the concepts, each is used where, and if, it is seen as aiding explanation. This last point illustrates how important it is to understand the different explanatory structures of physical geography.

Chapter 9 discusses the philosophical and conceptual basis of modelling in physical geography and different types of models are outlined. A range of issues associated with modelling are discussed including bounding, parameterisation, robustness and uncertainty as well as some of the steps taken by modellers to try to relieve these issues.

The final chapter, Chapter 10, tries to place the physical geographers themselves into context. The physical geographer is embedded within a series of social networks. These networks can influence what is studied, how it is studied, and the manner of its analysis. Beyond the Kuhnian view of researchers working within paradigms, or the Lakatosian image of research programmes, the social networks of physical geographers are much more varied and complicated. Individuals are not passive, mindless automatons following social trends, they are active agents working within different and often contradictory social networks. This more complex and subtle view of society (or rather societies) and physical geography is one that is relatively little explored. Likewise, the physical geographer is increasingly pressed by moral and ethical concerns within the research and wider community. Morals and ethics imply some form of responsibility, but to whom or what is the individual physical geographer responsible and why? Answers to these questions are not necessarily easy to come by, nor easy to discuss in a short textbook, but their potential importance in physical geography should not be underestimated. A perennial problem for geography of how, and whether, human and physical geography are linked and can remain an academic discipline is addressed, including recent developments in 'citizen' or 'participatory' science. There are no easy answers, as the discipline is not some detached and disembodied entity that is beyond the individual practitioner. Geography is a discipline because of the individuals within it; it is their practices and thoughts that determine whether it can remain united, or whether it divides. Is there anything more uncertain than how academics will act in the future?

Chapter 1

Ideas, change and stability in physical geography

Ideas are central to understanding how physical geographers have observed and analysed the physical world. Ideas are not static. Ideas change over time as well as from place to place. Ideas influence what we believe and how we understand reality. Ideas change internally, through the practice of a discipline, as well as externally through the social contexts within which physical geographers work. Differing ideologies, different philosophical approaches can influence the questions asked as well as the answers sought. Exploring this complex tangle of influences has often involved detailing the history of a subject. By following the 'great men' (and it usually was men in nineteenth-century academia) and their works, the key concepts from different periods can be traced and some sort of model of change developed. Often as important is the use of such a history to justify the current set of concepts within a subject.

This chapter briefly explores the history of physical geography to try to determine whether there is a coherent and accepted set of ideas that mould geographic thought. There may be no inevitability about the suite of ideas that form the focus of contemporary work. These ideas do not exist in isolation from particular philosophical approaches that have developed or have been imported into physical geography. Philosophies can hide in the background, directing and constraining the type of questions asked, and the way in which they may be answered. It is important to provide at least some illumination of these philosophical approaches and their relative importance at different times in and for different parts of physical geography.

What are ideas and how do they change?

Ideas and theories are frameworks of thought with which people try to understand physical reality. A more detailed definition of theory will be covered in Chapter 2, but for this chapter an appropriate definition of a theory would be the core ideas or concepts that go unquestioned when a researcher studies the physical environment. From this core, flow all the ideas and hypotheses that a researcher has about reality. The core ideas and the central relations between these and derived ideas form a tight, logical network. The dominance of vertical movement of land surfaces could be seen as an unquestioned assumption about how reality was in geology at the turn of the twentieth century. From this central core idea, other ideas flowed that explained the distribution of species that did not contradict the evidence on species distributions then available. Rapidly rising and submerging land bridges were mentally constructed to explain the range of distributions of fossil creatures between South America and Africa. As this central idea was replaced by the acceptance of horizontal movement, continental drift and plate tectonics were developed as an alternative explanation for these distributions.

A number of authors have tried to generalise how ideas change. Three 'big' ideas are: progressive change, paradigms (Kuhn, 1962, 1977) and research programmes (Lakatos, 1970). Progressive change states that ideas in science change incrementally and slowly. Current researchers build upon the work of others, correcting their errors until finally, at some vastly future unknown date, everything is understood and reality, as it really is, is known. Accumulation of establishable facts counts as progress of knowledge. Once a fact has been discovered it remains the same for all time. Current research is just adding to and extending existing knowledge (Figure 1.1). Unfortunately, understanding of reality does not seem to follow a nice, steady and progressive path. There are differences in the rate at which ideas change; some may suddenly appear apparently from nowhere and some seem to disappear, never to be mentioned in polite scientific circles again, such as racial superiority and acclimatisation. The view of progressive change assumes that earlier researchers understood reality and derived the fundamental laws and relationships that later workers have merely embellished.

Kuhnian paradigms were developed to explain how ideas in science appeared to undergo rapid and revolutionary changes. His prime example was the switch from the Greek to the Copernican model of planetary motion with the Sun replacing the Earth as the centre of the universe. Although much criticised, particularly by critical rationalists (such as Lakatos below), the idea of a paradigm has taken hold within geography and has been used to describe various perceived movements, fads or schools of thought. Paradigms, in other words, have been used to describe just about any group that is

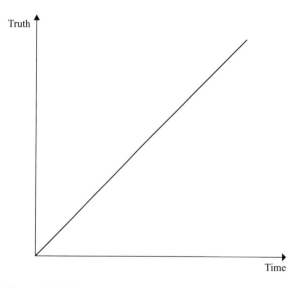

Figure 1.1 Science as a progressive pursuit of absolute truth.

perceived to have a common and coherent set of ideas about the world, whether that group realises they have these or not. Kuhn never really clarified or defined what a paradigm was and so there has been wide interpretation of what he meant by the term. He did outline what a paradigm consisted of. It was composed of a disciplinary matrix of tangibles and intangibles. Tangibles were examples of good or appropriate practice within a discipline: classic textbooks and classic experiments. These provided concrete illustrations of how to go about research and what was acceptable practice. In addition, tangibles included objects such as test tubes, laboratory benches and the like, which were viewed as essential to the practice of the discipline. Intangibles were the unwritten rules or conduct expected in a discipline, the social norms of behaviour and accepted evidence as dictated by the peer group of the discipline.

Change in this system was brought about not by the progressive and careful alteration of existing ideas in the light of new evidence or new techniques, but by the wholesale overthrow of old ideas by a complete and coherent set of new ideas. Kuhn suggested that most scientists undertake what he described as 'normal' science. They work diligently at their benches using standard techniques to prepare standard solutions to undertake standard tests. They ask only questions that they know they can answer using these techniques. There is little to challenge the established view of how reality is. This set of concepts and practices would make up the existing paradigm. Kuhn believed that facts or observations would slowly accumulate, reaching a point at which the existing concepts could not explain them. Eventually, the errors would become so pronounced, and would interfere so greatly with the operation of 'normal' science and so restrict the questions that could be answered, that there would be a complete switch from the existing ideas to a new set of ideas that explained the 'problem' facts (Figure 1.2).

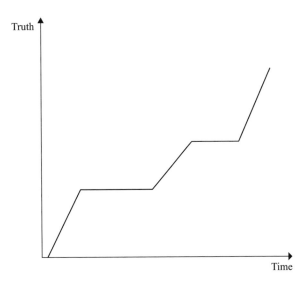

Figure 1.2 Science as a series of paradigms changing in a revolutionary manner towards a relative, but unknowable 'truth' – or at least that is what we would like to believe. In fact the lines could go in any direction relative to the absolute truth.

The paradigm switch was rapid and complete. 'Seeing' the world through one set of concepts, the paradigm, meant that you could not see it through the other set. There can only be one paradigm at any one time and seeing the world through that single paradigm excludes the possibility of all other views of the world. As with the gestalt diagrams though there is nothing to guarantee that your particular view of the image is the correct one. This is because switching from one paradigm to another means a complete change in what is viewed as evidence and even what is believed to exist in reality. This means that there can be no argument over the facts because the facts are different in each paradigm; indeed, what is viewed as a valid argument may even differ between paradigms. Paradigms need not operate across time alone. There may also be spatial differences in what is accepted as a valid view of reality.

How can we decide between paradigms? Kuhn argued that although it may be clear with hindsight that paradigm A was a better explanation of reality than paradigm B, at the time the choice could not be made on any rational basis. Instead, Kuhn turned to his set of intangibles and suggested that the choice between paradigms was based on sociology rather than 'hard', factual evidence. Peer pressure, rather than facts, decided which paradigm to accept. This was the part of the theory that critical rationalists found particularly hard to stomach. It implied that there was not necessarily any movement of ideas towards a true understanding of reality. If paradigm choice was socially based, then the new paradigm may not get you closer to how reality really is. Despite the new paradigm eliminating some of the errors or unexplained facts under the old paradigm, the improvement has been at the cost of all the concepts and understanding embodied within the old paradigm. There can be no rational discussion about the relative merits of either paradigm as there is no common basis for such a discussion. Kuhn's vision of science involved young, dynamic researchers overthrowing the old treasured and established views of an old guard. These researchers worked on critical experiments designed to disrupt the basis of the old order. Research into critical questions posed to undermine a paradigm was contrasted with the mundane 'normal' science of the established order. But does this romantic view of scientific progress match what actually happens?

Lakatos thought that Kuhn's ideas did not match his experience of science and how it changes. Lakatos was a strong supporter of a rational basis for choosing between competing sets of ideas and searched for a way to incorporate a critical rationalist approach into this process of choice. Lakatos divided sets of ideas into research programmes. Each programme had a central set of core ideas, the heart and soul of a set of theories. These directed much of the work the researcher undertook, constraining what questions to ask, how to ask them and the techniques and explanations that were valid in any answer (the positive heuristic). In addition, the research programme also identified the questions that researchers should not ask, techniques and modes of explanation that were unacceptable (the negative heuristic). The core theories were never directly tested, they were kept away from direct assessment. Instead, a band of auxiliary hypotheses were the basis of the active use of each research programme. These were testable statements derived from the core theories. These were the elements of the research programme that researchers used in their scientific lives. Disproving one hypothesis did not bring the whole of the research programme into question; instead, that hypothesis was rejected and some other hypothesis constructed using the existing central theories (Figure 1.3).

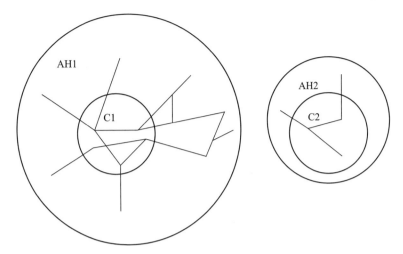

Figure 1.3 Lakatosian research programmes. A central core of theory (C1 and C2) surrounded by a protective belt of auxiliary hypotheses (AH1 and AH2). C1 and AH1 represent a progressive programme expanding and generating new hypotheses. C2 and AH2 represent a degenerative programme, with a belt of auxiliary hypotheses contracting towards the core. The nodes represent axioms in the core and hypotheses in the outer circle. These nodes are linked by vertices that represent relationships between nodes. The progressive research programme has a rich and internally highly connected core of axioms. These generate a range of hypotheses that can even interact with each other to generate 'second'-generation hypotheses (or more). The degenerative research programme has a limited set of core axioms that generate few isolated hypotheses.

Lakatos suggested that choosing between different research programmes became a matter of rational choice. Research programmes that continually generated new hypotheses, novel ways of questioning reality and which were generally not disproved were progressive programmes. Research programmes that were stagnant and did not advance new ideas or explanations for reality were seen as degenerative. When there were two competing research programmes researchers switch to the more progressive research programme. Some individuals might take longer than others to see that one programme was more progressive, but eventually all would make the same rational choice. In Lakatos's view of science there was elimination of errors and change of world view, but each change brought the researcher, or rather the scientific community, closer to the truth, to the way reality really is. Unfortunately, it would never be possible to prove that reality had been captured absolutely by the theory, only that the theories of a particular programme had not been disproved.

A further view of how ideas change is provided by Feyerband (1975, 1978). Feyerband's argument is that there is no single, scientific method and that in fact science is an anarchistic enterprise. He suggests that the only principle that does not inhibit progress in science is that 'anything goes'. He points out that even ancient and seemingly absurd ideas can help enhance our knowledge of reality. As with Kuhn he noted that not all facts fit with contemporary theories, but unlike Kuhn he viewed this as illustrating that 'facts' were a construct of old ways of thinking and that the clash between fact and theory highlighted the poverty of trying to develop overarching theoretical frameworks or paradigms. But Feyerband also criticises falsification as a basis for a special method in science, as falsifying a theory or idea will result in falsifying away all of scientific thought associated with that theory or idea. Feyerband points out that science is viewed as somehow different from other explanatory frameworks because it appears, through technological success amongst other things, to have a special method for understanding reality. He believes that this is an illusion.

> There is no special method that guarantees success or makes it probable. Scientists do not solve problems because they possess a magic wand – methodology, or theory of rationality – but because they have studied a problem for a long time, because they know the situation fairly well, because they are not too dumb . . ., and because the excesses of one scientific school are almost always balanced by the excesses of some other school.
>
> (Feyerband, 1975, p. 8)

Feyerband's views, together with those of Kuhn, are useful in identifying the context of scientific discovery and analysis, and in bringing to the fore the importance of sceptical thought even if it is through the use of seemingly absurd ideas. Viewing science as being as much of a fairy tale as any other explanatory framework is, however, a criticism too far for most scientists. Science may be a human construct but the approach scientists have taken is always open to systematic criticism. And, whatever anyone says, success in terms of seeming to explain reality and to manipulate reality using these explanations does imply some coherence between what scientists think and how reality is, even if we can never be sure what that coherence actually is.

Both paradigms and research programmes have been used to describe changing concepts in physical geography. Neither is perfect and each has a number of problems. Adhering to paradigms would imply that at any one time a single, all-embracing set of ideas would direct the researchers in a subject. This is rarely the case. Similarly, adherence to research programmes implies that central sets of concepts are somehow immune from testing and unaffected by the failure of hypotheses derived from them. Such an ivory tower existence for privileged theories is rarely sustained. Some authors, such as Stoddart (1981), have suggested that there are general currents in thought and that theories follow these. Researchers are moved by fads and fashions as much as anyone else and the flavour of the month has as much relevance in describing academic work as any other. Identifying how such fads are translated into theories, or a mechanism by which changing ideas occur within fads, has not really been tackled. A significant problem with both frameworks is that they do not really address how and why certain ideas seem to survive and prevail, whilst other ideas are rejected. The stability of some concepts, the recognised, but malleable nature of others, and the outright rejection of still others, is not addressed.

How ideas change is still open to debate despite the models presented here. One of the important points to bear in mind is how each model views the ultimate goal of research. If you believe that knowledge is progressive and developing towards a final complete understanding of reality as it is, then Kuhn's paradigms with their sociological basis are unlikely to be attractive. If you believe that all knowledge is relative, then paradigms may be more attractive. Lakatos seems to sit on the fence by accepting that ultimate absolute knowledge is unattainable but then implies that there is a progressive march to this goal that helps to decide between research programmes. This would suggest an underlying belief in the movement of ideas towards representing and explaining reality as it really is, even if you can never be certain that you have arrived.

Johannes Umbgrove and plate tectonics

Johannes Umbgrove (1899–1954) was a Dutch geologist working at the cusp point between belief and disbelief in plate tectonics. Most histories of plate tectonics are written from the viewpoint of the protagonists trying to establish the 'new' idea (e.g. Wood, 1985 or Oreskes, 2003) with Alfred Wegner often portrayed as the tragic genius who foresaw the whole theory. Umbgrove is a different character. He represents the geologists trying to grapple with the inconsistencies of the contemporary theories without recourse to the geophysical-inappropriate theories of continental drift. Umbgrove was not a minor figure in geology at the time. He undertook interdisciplinary research into gravitational surveys, palaeontology as well as tectonics and was amongst the first geologists to view the earth as a single dynamic system. Umbgrove, like most geologists at the time, had identified key episodes within different geological periods of simultaneous development of topographic features such as mountain chains across the globe. He referred to these periods of activity as 'pulses' and they formed the basis of his 1947 book *The Pulse of the Earth*. Within this book he correlated episodes of orogenesis (mountain building) with other geologic phenomena such as sea-level changes, magmatic cycles and ice ages. He viewed these 'pulses' of activity as interconnected and in search of a common, overarching explanation.

Umbgrove provided a thoughtful consideration of the competing theories for this common cause (Umbgrove, 1946). One theory, for example, put forward the idea of the submergence of large continental blocks as a means of producing the current distribution of continents and oceans, whilst another theory suggested that continental blocks had extended to produce this observed distribution. A more complicated theory viewed all of the crust as having been originally simatic rock (silica- and magnesium-rich) and erosion had removed the basic part of this rock from the continents and so formed a sialic continental crust (silica- and aluminium-rich). These continental crusts are less dense than the oceanic crusts and so, in restoring isostatic balance, these portions of the crust rose and the simatic areas buckled and subsided to form the oceanic basins.

Umbgrove came up with his own alternative explanation. He saw the early Earth as being coated with a sialic layer of rock that slowly solidified and floated on a denser substratum. Convective currents in this denser substratum caused the crust to buckle, fold and drift to produce a pattern of sialic-rich and sialic-poor regions of crust. This convective activity was, however, limited to the pre-Cambrian. After a critical point in this long geological era, horizontal movement ceased as the mantle become too rigid to permit such movement. The distribution of continents and oceans became frozen.

> Such youthful capacities . . . may be expected to have ended as soon as the earth attained a physical state more or less resembling that which it has at present. From that time onward the terrestrial forces were imprisoned and became operative as subcrustal processes.
>
> (Umbgrove, 1946, p. 252)

In advancing his theory, Umbgrove was not contradicting the prevailing view amongst physicists, in particular Harold Jeffreys (1952 and as late as 1970), that the earth's mantle was too rigid to permit horizontal movement of crust. His theory does explain the observed distribution of mountain chains and a whole range of other morphological and geological evidence that he had also used to counter Wegner's continental drift theory.

How does Umbgrove's theory help us to understand how science works? His theory uses concepts such as 'convective currents' that were to become essential as causal mechanisms in plate tectonics, so it is clear he understood the geological thought of his time. His theory seems, from our distance of almost 70 years, like the compromise that it is. This does not mean it is bad. It means that Umbgrove tried to reconcile the observations of patterns with the accepted conceptual tools available to him. Umbgrove's theory illustrates that any scientific explanation does not stand alone, it draws upon and gains strength or credibility from the existing body of scientific thought and, where possible, tries not to contradict the more 'certain' facts within this body of thought. In Umbgrove's case this meant reconciling horizontal movement with the 'fact' of the rigidity of the earth's mantle, a problem he circumvented by hypothesising that this rigidity did not occur until a certain point in the pre-Cambrian, thus allowing some horizontal movement of crust.

Is there a history of ideas in physical geography?

Physical geography is a diverse subject with many evolving subdisciplines vying for status. Outlining a single history of such diversity is both problematic and misleading. Detailed histories of ideas and their development can fall into the trap of reading the past by the light of the present. Trends, unbeknown to workers at the time, are identified with contemporary certainty and clarity. Such histories can lapse into justifying the present rather than trying to understand the past concepts in their appropriate contexts. Often these histories are involved in justifying current practice or concepts by referring to a long and hitherto neglected (or unknown) history. Such reinterpretations of the past often provide intellectual justification for the present by an appeal to the authority figures of a noble past.

The first thing to note in looking at the history of physical geography is that there is not *a* history of physical geography. There are many possible histories that can be constructed from the information available. Judgement of a single narrative as the 'correct' one implies a very restricted view of what is important in physical geography and how it came to be. If there are many possible histories, how can anyone identify what is important from the past? It is essential to have a clear idea of why you want to look at the history of physical geography. This objective will determine what you consider to be important. In this book, our purpose in reviewing the history of physical geography is not to provide a definite history. Instead, there are three key concerns that we want to highlight as being present in the study of the physical environment: universality of explanation, a concern with the empirical and a concern with explaining change or stability. Tracing how these concerns have been expressed and the strategies for answering them is the focus of our brief review of history. Even this limited ambition is fraught with difficulties.

Delving into the history of physical geography requires identifying the subject in the past. This can be a major problem. Individuals who identified themselves as, or what they were doing as, physical geography requires the identification of physical geography as a legitimate subject area. As an academic subject, physical geography requires professional academics, academic institutions within which they work, journals within which papers in the field can be published, and so on. The subject requires extensive legitimising networks of relations between academics, between institutions and other organisations such as publishers. These networks were not necessarily around in the past nor in a form that would permit the identification of physical geography as understood today.

A more fruitful approach might be in identifying the central subject matter and defining the histories of physical geography that way. There is a problem with this approach as well. It assumes that the same subject matter has been studied throughout academic disciplines over time and space. Although it may be possible to identify the ancient Greeks as studying climate, were the phenomena they studied the same as what contemporary or even Victorian academics would consider climate? Were they collecting information on the same phenomena? Even if there was some commonality in the nature of the phenomena under investigation, is it possible to abstract it from the whole intellectual framework of which it was a part? Newtonian mechanics is usually viewed as a key element of the foundation of modern science and Newton's 'modern' approach of experiment confirmation is viewed as exemplary. The alchemy aspects of Newton's work, and the manner in which this was intertwined with his 'acceptable' science, is rarely mentioned (White, 1998). For Newton, both aspects of his work were vital; he made no distinction between what we would deem 'proper' science and 'psuedo-science'.

Another possible approach is to look at the 'great' figures in the subject and define your histories by reference to what they thought and did. Although it may not be possible to identify them as physical geographers (remember the term might not have existed), at least their work survives to be reinterpreted as the appropriate subject matter of contemporary physical geography. Appropriating figures such as Darwin, Wegner and Huxley as influential in developing ideas in physical geography neglects the multi or even interdisciplinary nature of their work. The environment these individuals worked within was different from the present and the subject areas they recognised would not have necessarily included what contemporaries would view as the 'core' of physical geography. It is also likely that other disciplines will claim these figures as originators in their subjects as well. Appropriation of 'great' men will involve teasing their relevant work from the rest of their intellectual endeavours, a division they would not have made. It could be argued that even this approach creates histories that mirror contemporary concerns, rather than really reflect the work of these individuals. Likewise, focusing on great men (and it usually is men in the gender-closed Victorian intellectual world of geography), neglects the contribution of 'lesser' figures who undertook practical work that might be described as geographical. The army of surveyors, for example, who undertook the surveying of the British Empire, trained in and developed the methods for describing and collating spatially referenced information (Driver, 1992; Collier and Inkpen, 2003). The data and analysis undertaken by these forgotten figures was vital in developing imperial policies and research methods, yet it is often regarded as a lesser achievement than abstract theorising. Contemporary intellectual snobbery can cloud what is deemed relevant and what is deemed unworthy for a subject's history.

Most anglophone histories of physical geography tend to focus on the Anglo-American experience and definition of the subject. This limits the types of histories that can be written. Although it could be argued that this impression reflects an inability to read well in other languages, it also reflects a belief in these areas that they represent 'real' geographic thought. Some histories may acknowledge the early work by Portuguese navigators or Arab intellectuals, but from the eighteenth century onwards the Enlightenment and scientific revolutions bring with them a view of 'proper' science finally being done. For proper, read science as we, Western Europeans, would understand it today. Such views are also translated into contemporary global surveys of geography. Anglo-American derived topics and ideas occupy centre stage, other cultures are relegated to minor interests. It could be argued that this concern with Anglo-American physical geography reflects a wider pattern of change since the eighteenth century. Development of a capitalist world economic system, some would argue, has marginalised and stifled any independent intellectual development of indigenous cultures. It may be, therefore, not surprising that the only 'acceptable' histories are those with a Western, and specifically an Anglo-American, flavour, although acceptable to whom and why is another loaded question that is too complicated for consideration here. The possible bias in the construction of histories, and the reasons for it, are rarely touched upon in histories of physical geography.

All the above might imply that there is no point in trying to even produce an historical view of physical geography. There is no single history, there is always the problem of interpretation and there is always likely to be subjective bias. There is, however, still a case to be made for sketching even a flawed history of physical geography. Identifying ideas within their context and tracing how some ideas have evolved may help in understanding why some ideas remain relatively stable and others fade into obscurity. Stable ideas are those that appear to retain their basic tenets and produce results that are acceptable to the academic community of the time. The Davisian cycle of erosion, for example, could be viewed as an idea that no longer provides acceptable results upon which contemporary geomorphology can operate. It does not provide a framework for asking or answering questions felt to be appropriate to practising geomorphologists. The cycle of erosion becomes a distinguished or obscure (depending on your viewpoint) element of the history of the discipline. Plate tectonics, on the other hand, has evolved to be a major and overarching framework for posing and answering geomorphic questions about long-term landscape change. From the 1960s, the general idea of plate tectonics has been developed and refined to create a complicated range of subdisciplines, each with their own particular criteria for evidence and working practices. Tracing the decline of one idea and the rise of another brings to the fore both the shifting nature of what is viewed as real and the surprising stability of much of the content of physical reality.

What are the important concepts in physical geography?

The three themes developed as central to the history of physical geography: universality of explanation, a concern with the empirical and a concern with explaining change or stability, are not independent. The quest for universality in explanation has been driven by the need to explain landscapes and their changes. Likewise, identification of the changing or stable nature of the physical environment has been aided by increasingly complex empirical information. The search for universality can be viewed in terms of both an overarching theory as well as an integrated method for analysing the physical environment. Huxley's *Physiography* (1877) is viewed by Stoddart (1975) as an important exercise in forming an integrated approach to the subject matter of the physical environment. Stoddart views Huxley's approach as being in a direct lineage from the conception of the physical environment of Kant and Humboldt nearly a century before. Physiography starts at the microscale with the familiar, and works outwards to the macroscale and unfamiliar. In so doing, Huxley develops an explanatory framework that begins with identifying and classifying causes at the local level and then working from these familiar, or as he would have it 'commonsense' illustrations, to the wider picture. Cause and effect relationships are built from individual experience, emphasising the empirical over the theoretical, but applied to increasingly larger spatial and temporal scales. The local is then viewed as part of this wider context, and understanding of the local requires this wider context.

Huxley's approach had much in common with the 'new' geography being developed by Mackinder (1887) in highlighting the importance and significance of integration and synthesis of information, both physical and human. Much as with Mackinder's new geography, the high ambitions of physiography were never really fulfilled in developing physical geography. Davisian geomorphology overtook Huxley's physiography as the main explanatory framework of geomorphology and the dream of an integrated study of the lithosphere, atmosphere and biosphere was increasingly replaced by specialisations.

From the perspective of the early twenty-first century, the failure of both Huxley and Mackinder to have the influence that their holistic and integrative approaches should, or were supposed to, have is of interest. These two figures were influential in their own right and yet their ideas, which strike several chords of recognition, failed to develop as the core of the subject. This illustrates the first problem in outlining a history of physical geography. Both Huxley and Mackinder are viewed as using a Darwinian framework for their ideas, as being good disciples of Darwinian thought. Does their failure then reflect a failure of Darwinian thought to influence physical geography? According to Stoddart (1966) the answer would clearly be no; Darwin's views of ecosystems, the integrative nature of the environment and the significance of change, have permeated geographical thought. The central core of Darwinian thinking (Gould, 2000), that of non-directional evolution through random variations in organisms, has had a lesser impact upon geographical thinking. It is only specific interpretations of Darwinian evolution that have affected geography, not the detailed theory as outlined by Darwin in 1859. The first problem then is not to fall into the trap of assuming that there existed in the past a single or universally accepted interpretation of a potentially integrating concept, such as evolution.

A vital rallying point in the search for universality in explanation, as noted by Goodman (1967), was the concept of uniformitarianism. The term, usually attributed to Charles Lyell (Lyell, 1833), has been paraphrased as the idea that 'the present is the key to the past'. The general idea is that by observing the present in terms of forms and processes, it should be possible to apply this knowledge to the past. Everything that happens now should have happened in the past. Gould (1965), amongst others, has pointed out the major problems with maintaining a substantive rather than a methodological view of uniformitarianism. Gould (1965) defined substantive uniformitarianism as the concept that the rates of operation of geological processes have been constant or uniform throughout time and space. Upholding this view would imply, for example, that there has been constancy in the number and type of events causing the landscape to change and so in the magnitude and frequency of the processes causing change. This view of uniformity led Lyell to an extreme cyclical view of landscape and biological change (Gould, 1987; Kennedy, 1992) in which he could seriously envisage a cycle to life with extinct species eventually being recreated. Methodological uniformitarianism referred to the assumptions that underlie any study of the past and present in an historical and empirically based science. First, the laws that operate to produce change or stasis are assumed to be constant in both space and time. This is a vital assumption as it means that explanations for change and stasis in one location can be applied to other locations and other time periods with the confidence that they will operate in a similar manner.

Second, it is assumed that explanation of the present (and indeed the past) does not require any invocation of unknown or unknowable processes. This is the assumption that uniformitarianism attacked most saliently. At the time of its development, uniformitarianism was a concept set up in opposition to the prevailing attempt to explain all of nature, and more importantly what was not understood about nature, by final reference to an overarching deity. This final source of all explanation could be invoked to explain why the world was as it was; there was no need to seek a material, Earth-based explanation – reference to the guiding principles of a deity was sufficient. This should not, however, be interpreted as meaning that the explanations offered were simplistic or unreasoned. On the contrary, the most learned brains of the time subscribed to this view and debated the 'correct' interpretation of evidence in a logical and reasoned manner. It is just that the basis of their logical system permitted explanation by reference to a deity as much as explanation by reference to observable processes and agents. Hutton and Lyell used purely material, empirical observations and reasoning restricted to what they could observe and infer about material reality to construct explanations of features in the landscape. This was a radical step and at a stroke it removed catastrophism from scientific explanation. This was important as it meant that dramatic, biblical events, such as the flood, could no longer be used to explain observations such as extinctions or the unconformity of sediments. Likewise, it meant that observations no longer had to be shoehorned into a temporal sequence that was restricted by biblical events. Acceptance of the glacial theory of Agassiz, for example, required the acceptance of an Earth considerably older than 4004 BC, as estimated in 1658 by Archbishop Ussher of Armagh. In this context, the presence of large boulders that scattered the landscapes of North America and Northern Europe could be explained by the slow process of glacial advancement and retreat, rather than having to be explained by a biblical flood in order to fit the constraints of a Young Earth. Bernard Freiderich Kuhn, in 1787, interpreted the presence of large boulders in the Jura mountains of Switzerland as evidence of an ancient glaciation, an interpretation subsequently supported by Hutton in 1794 after visiting the region. Evidence and support for an ancient glaciation grew following work by Esmark in Norway, Bernhardi in Germany and de Charpentier in Switzerland, and then Agassiz himself, who subsequently assimilated the available evidence and formulated a coherent theory of 'the great ice period', in 1837, and wider acceptance of the glacial theory gradually followed. Given the societal context within which the glacial theory of Agassiz developed, one in which religious belief permeated and dictated the remit of scientific inquiry and interpretation of phenomena, it perhaps comes as no surprise that the broad acceptance of the glacial theory took many years.

The empirical 'facts' now had greater weight in a scientific argument than the enforced explanatory framework. Time, as the arena within which processes operated, was no longer limited to the scale of biblical events. Out of this release, however, another constraint was developed – the idea that there could be no sudden change in processes, or the occurrence of processes, not currently active. This meant that explanations that made use of what would today be considered as extreme events, such as meteorite impacts, were excluded from the explanatory framework.

The important ideas of uniformitarianism lie in the concepts from which it is derived. Uniformitarianism assumes that everything observed in nature can be explained, the principle of causality as an overriding universal principle enters into explanation. Everything observed is capable of explanation, everything observed has a cause. The cause is not based in some unobservable and hypothetical other world, in some omnipotent deity. Causes are to be found in what we can observe in the here and now. Cause and effect only have recourse to the material world, not to the supernatural. Uniformitarianism is in this sense, as Goodman (1967) suggests, the simplest explanation for the physical world and how it changes. From this one idea, applied consistently and continuously in physical geography, the importance of the empirical rather than the theoretical, of the observable rather than the unobservable, and of the location of causality in the material present are to be found.

Of vital importance in nineteenth- and early twentieth-century thinking about the physical environment was the use, or rather interpretation, of the concept of Darwinian evolution. Darwinian evolution could be seen as representing the application of uniformitarian principles to the organic realm. In this manner, it is another attempt to provide an overarching theory to explain the natural world. Evolution is, however, more than just Lyell amongst flora and fauna. Bowler (1983a,b) has written on the development of the concept of evolution before and after Darwin, and Livingstone (1984) has illustrated how the concept of evolution was interpreted in a particular manner by American physical scientists. Darwin had expounded a theory of change in the organic realm by means of random variations in individuals. Individuals competed for resources and, where the random variation was beneficial in this competition, successful individuals produced more offspring. The offspring had the beneficial variation that became a dominant characteristic in the species over time. From initially small differences between individuals multiplied over the vastness of geologic time, species became differentiated. The random variation upon which this process of natural selection was based was non-directional.

This means that the course evolution takes is not predictable; uncertainty and randomness lies at the heart of Darwinian evolution.

Central to North American interpretations of evolutionary thinking within physical geography was the interpretation of evolution used by W.M. Davis in his cycle of erosion. Essentially Davis used the concept of progressive change through predefined stages as the basis for his view of evolution. This view was not the one expounded by Darwin involving random variation and non-directionality. Davis's view, however, was firmly grounded within a North American tradition of orthogenesis and progressive change as developed by Hyatt and, within Europe, by Haeckel. Haeckel viewed changes in evolution as occurring by the addition of stages to an existing established sequence. An embryo, in his view, went through all previous evolutionary stages, with a new stage, that of humans added at the end of the embryo's development. The same sequence of stages could be observed in embryos of different species, but for the 'lower' species this happened without the addition of more 'advanced' stages. So whilst a human embryo might be expected to go through all stages from fish through amphibians to reptiles to mammals, a fish would not have these additional stages visible in its embryonic development.

Gould (1977) uses the term 'terminal addition' to describe Haeckel's vision of evolution. Gould also notes that this viewpoint was compatible with the Lamarckian view of the inheritance of acquired characteristics and that Haeckel did indeed subscribe to the transfer of such characteristics between generations. Hacekel's views from biology have a resonance with the ideas of Cope (1887) and Hyatt (1897), the latter a palaeontologist working initially under Agassiz in the US. They believed that although there may be a preordained sequence of stages, organisms could acquire characteristics as well and pass these on through generations. Central to Cope and Hyatt's work was the idea that the stages that organisms went through in their development could be accelerated or slowed down. Progressive evolution saw stages being condensed further and further into the early stages of development, allowing more time for the addition of acquired characteristics at the terminus of development. Retrogressive evolution saw stages being slowed almost to a stop so that stages are lost before development ends. The former they termed the law of acceleration, the latter the law of retardation. Figure 1.4 illustrates these two ideas. Their significance lies in the possible influence they had upon the development of evolutionary ideas in the US. Rather than the non-directional, random and chancy process of selection for natural variations in characteristics of an organism independently in each organism's lifetime, Cope and Hyatt strove for an almost opposite view of change. Within their scheme, evolution was about movement through a series of predefined stages with acceleration or retardation adding or losing stages through the sequence. This is the form of evolution that should be borne in mind when discussing Davisian evolution and the development of ideas from it.

Any landform and any landscape had to pass through the stages set by Davis in his cycle of erosion. Any landform could be classified by its morphology to a particular stage in this predetermined evolution, through youth, maturity and old age. Explanation was contained within the theory, within the evolutionary model used. Observation merely confirmed the position of a specific landscape or landform within this scheme. Stage became the explanation rather than what happened in the landscape. Significantly,

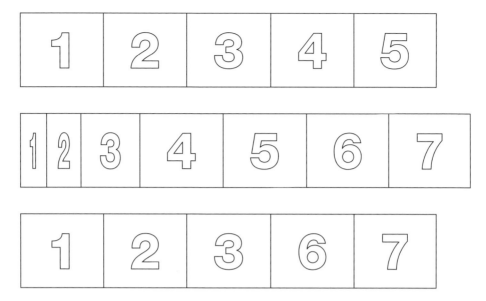

Figure 1.4 Initial development sequence of even periods of change. Acceleration results in contraction of time periods 1–3, but periods 6 and 7 are added roughly within the same overall development span. Subtraction results in periods 6 and 7 being added within the total development span, but periods 4 and 5 are removed.

however, the stage model highlighted the importance of identifying and providing an explanation for stability and change in the landscape. Both types of behaviour could be explained by reference to the same explanatory framework. Change was an essential feature of evolution and stability was illusory, based on the inability of humans to perceive alteration at the timescale of relevance to the landscape.

These characteristics of the cycle of erosion provided it with a scientific veneer, particularly as it explicitly used evolution, the scientific theory par excellence in the nineteenth century. Bishop (1980) and Haines-Young and Petch (1986) have pointed out the deficiencies of the Davisian model as a scientific theory, but its perception as a source of progress in explanation in physical geography in the nineteenth century and as a guiding framework for geomorphic thought in that period is generally not questioned.

Davis's interest was, however, focused at the macrolevel only, although his quest for explanatory frameworks admitted to the operation of the same processes, the same models, across all scales as in his work on tornadoes and their similarity to the spiralling of galaxies. In this manner, Davis believed he had found an overarching and universal mode of explanation for phenomena in the physical environment. It would be inappropriate, however, to view nineteenth-century physical geography as just about Davis and geomorphology.

Even at the scale of the landscape, however, the Davisian model was not the only available theory – Penck also developed a model of landscape development. Penck's model has been reinterpreted for the systems era as an early attempt to apply a coherent systems framework to the diverse factors affecting landscape development. Although Penck would not have understood his theory in those terms, his use of factors and their relations to define landscape development does seem capable of easy translation (e.g. Thorn, 1988). A significant difference between Penck and Davis was the nature of changing slope form in each theory. In Davis, slopes declined, whilst in Penck slopes were replaced. In slope decline, denudation processes remove material from the top of a slope at a faster rate than they remove material from the base of the slope. Gradually, the slope as a whole declines in angle. In Penck's theory, the base of the slope becomes covered with debris denuded from the top of the slope. Gradually, the base of the slope is replaced by a debris slope of constant angle. In the 1950s, King proposed a third type of change, slope replacement. In this theory, slope elements remained constant (waxing, waning, free-face and pediment); only the length of the pediment slope increased over time as debris from the other slope elements reached it and extended it. All other slope elements remained at the same angle throughout their existence. Although differences in the three theories can be partly related to the different locations of their production, as noted by Thorn (1988), the three theories do illustrate a commonality in their use of time. All three are time-dependent, i.e. all three presume that change will be progressive over time and the nature of the form of the landscape will change over time.

Denudation chronology has been viewed as a mode of explanation directly linked with the Davisian cycle of erosion and often touted as the dominant mode of explanation in physical geography until the late 1950s. Denudation chronology attempted to reconstruct the past evolution of the landscape from the sequence of denudation surfaces still visible in the landscape. Evidence for this relative dating of land surfaces was to be found in the morphology of the landscape, its sediments, from absolute dating techniques of portions of the landscape and from an understanding of processes of landscape development (Gregory, 1985). Any evidence collected was assessed within a framework of stages of development of the landscape. Evidence did not stand alone but had to be interpreted within this framework. A great deal of the methodology and theoretical debate over the identification and presence of entities such as planation surfaces centred, as Gregory (1985) noted, upon a few 'classic' studies such as Wooldridge and Linton's (1933) inter-pretation of the South Downs. Even the advent of statistical analysis of data did not initially diminish the appeal of this approach to landscape study, as trend surface analysis was employed to aid the identification of 'real' surfaces in landscapes.

Another alternative set of theories concerning landscapes and landforms presumed that change could be understood as time independent. In these theories, forms did not change over time, there was instead a steady balance between variables that produced forms that could not be placed in a temporal sequence. Gilbert's concept of dynamic equilibrium and Hacks's model of the same are classic examples of this approach. Both Gilbert and Hack viewed spatial variations as more important than temporal variations. Spatial variability replaced temporal variability as the dominant feature of these theories. In this view, the landscape is in balance with current inputs and so there is no temporal memory in the landscape. The only possible source of variation in the landscape is not in the different stages in development or even in the past histories, but instead, lies in the spatial variation in nature and inputs into the landscape. Although the exclusion of the 'past' as a source of variation in the landscape is a problem in time independent concepts, they do form a basis for trying to understand the landscape as a product of contemporary processes. These models bring landscape-forming processes to the time and space scale of the investigators and detach the landscapes from their unknowable (in human terms) pasts. They provide, in other words, a platform for examining the landscape at spatial and temporal scales relevant for applications such as engineering.

The above two approaches to landscape study are very Anglo-American in their orientation. There existed other theoretical frameworks within which landscapes could be explained. Two of the most significant in continental Europe were climatic and structural geomorphology. Climatic geomorphology, as the name suggests, viewed climate as the overarching explanatory framework for landscapes. This is the view that climate controlled the characteristics and distribution of landforms. Climate is the controlling factor in landscape development. Climate resulted in a distinct latitudinal zonation of landforms with an additional similar variation for the effect of altitude. There was a consistent causal relationship between the two, between climate and landscape, based on the persistent and consistent presence of a set of land-forming processes unique to each climatic environment. A zonation of landforms on the basis of climate was also developed by Tricart and Cailleux (1972) in France. The climatic zones, and indeed continents, were not static and this was reflected in the landforms found in different regions. Landforms and landscape could reflect the action of past climates, and so the simple latitudinal zonation had superimposed 'historical' patterns of climatic influence. Linton's (1955) model of tor development on Dartmoor was a classic example of the application of climate and its change as the explanation for the development of a specific landform. Linton hypothesised a two-stage climatically controlled series of processes. First, the core stones of the tors were developed through the deep weathering of granite under tropical conditions. Large core stones were produced at depth where weathering was less intense and smaller core stones near the surface where alteration was greatest. The regolith remained around the core stones until removed by an episode of periglacial activity. This climate was more effective at removing weathered material and, with this regolith removed, the core stones formed disjointed masses on hilltops and valley sides. The idea of a 'natural' zonation of the physical environment, particularly as controlled by climate, was not only applied to geomorphology. Herbertson (1905) developed his idea of natural regions based on the distribution of biota under the influence of climate on an ideal continent.

Peltier's (1950) version of the importance of climate related more to the relationship between process and climate than directly to landforms. Peltier insisted that climate controlled the ability of processes to operate. A particular combination of temperature and precipitation would invariably result in the operation of mechanical or chemical processes of denudation or a particular combination of both. Although this vision of geomorphology was developed, and still influences geomorphology on continental Europe, it found little favour in the UK as noted by Gregory (1985, quoting Stoddart, 1968, Derbyshire, 1973 and Douglas, 1980). The main criticism seems to have been that the idea of climate affecting landform and landscape development was perfectly acceptable, but the idea that it was the major theory for landscape development was not. The view of a specific climate resulting in a specific landscape was seen as too simplistic and not able to cope with the complexities of interpretation that structural differences and past histories imposed on it.

These early trends in classification and quantification of natural phenomenon tend not to be mentioned in 'standard' histories of physical geography. The reasons why are unclear.

One possible reason is that these measurements were often made without the use of an overarching explanatory framework within which to place the measurements. Gregory (1985), for example, notes that Kuchler (1954) categorised the early twentieth century as a period of data collection and accumulation during which there was no co-ordinating or overarching theory with which to explain the patterns observed. Within subdisciplines such as meteorology, however, this was not the case as theories of atmospheric circulation and weather patterns were available to co-ordinate and explain temperature patterns, windflows and other climatic parameters. Solomon (2001), for example, illustrates that there were accepted explanatory frameworks based on detailed theories for weather patterns even in the data-poor Antarctic region at the end of the nineteenth century. These theories were used to predict the likely behaviour of weather patterns and in expedition planning, although the extreme and unpredicted weather of 1912 resulted in the tragedy of Scott's failed expedition. Although the early twentieth century would see these patterns and classifications extended by the likes of Thornwaite (1948) and Koppen (1900), the numerical basis for these schemes was developed in the nineteenth century.

By the 1960s, the questions that physical geography was asking of the environment had altered, and the explanatory frameworks used had to evolve as well. Davis's cycle of erosion focused research onto the megascale, onto questions of landscape development over aeons. Pressing questions of engineering concern, such as how is a hillslope going to behave if a road is cut through it, were not answerable within this theoretical framework. Whether this type of question was appropriate for this framework, and by implication an appropriate question for geomorphology, was not, it seems, a concern. The nature of questions in physical geography had altered since Davis posed questions on a 'grand' scale. The quest for relevance and the need to justify their academic existence meant a focusing of geomorphology on questions of relevance to society. Whether these questions were questions of relevance to everyone, or to particular groups in particular locations, has not been made clear in the histories of the subject. Of central concern was, however, the recognition that most questions concerning landscape on a human, spatial, and temporal scale were unanswerable using the Davisian framework. Identifying a planation surface did not help in identifying the stability of a specific hillslope, nor did the identification of a section of the landscape as mature help in identifying and predicting rates of change. The nature of questions of interest to geomorphologists had altered and so their operating framework had to change.

Other parts of physical geography had long discussed issues at scales other than the regional and geologic. Climate studies, for example, had integrated short-term measurements of changes in variables such as temperature into explanatory frameworks at a range of scales. Predictions of weather patterns relied on the rapid collection and processing of information within changing models of atmospheric operation. Small-scale and large-scale phenomenon were both recognised and modelled using the same physical laws. Aggregation of behaviour was needed to predict at the regional and global scale, but the principles on which the models were based were similar. Ecological studies had likewise developed a scale-independent framework for analysing and integrating information. The ecosystem concept had been forcefully put forward by Tansley (1935). The ability to identify, classify, bound and relate entities within the landscape at any scale made this form of explanation an extremely powerful tool for understanding.

Geomorphology overcame the scale problem of the Davisian model by sidestepping it. As Haines-Young and Petch (1986) noted, the solution was not a new theory, it was an alternative means of interpreting the information that already existed. The new interpretation could not be tested directly, but its explanatory framework offered a means of ignoring the temporal and spatial scales of Davisian focus. Schumm and Lichty's classic paper (1965), discussed further in Chapter 2, overcame the Davisian framework by restricting its scope of operation to a specific type of time, cyclical time (Figure 1.5). This was the longest temporal scale they envisaged. At this scale only certain variables, as identified by Davis, were of significance in explaining landscape development. In other words, there were only a limited number of variables that changed at this scale. There were, however, two other temporal scales, steady-state and dynamic equilibrium. At these two scales, the variables that were significant for long-term landscape development could be treated as if they were constant and so were unchanging. Other variables, considered insignificant in cyclical time, became variable, and so significant, at the other two temporal scales. In this manner Schumm and Lichty provided the beginnings of an explanatory framework that retained the Davisian view, but which permitted other conceptual frameworks for explanation to be applied at other temporal, and by implication, spatial scales.

Systems analysis (see Chapter 7) developed rapidly as the integrative explanatory framework of physical geography. As important, however, were the 'new' questions that this approach made legitimate, questions already being asked and answered about processes of change and their rates. Although the importance of process for understanding geomorphology had been recognised (e.g. Strahler, 1952) and even attempted as a basis for classification in climatic geomorphology, the seemingly immature development of the subject relative to the 'hard' sciences seemed to condemn geomorphology to 'mere' description of forms. Identifying and understanding process and their rates of operation was essential, even to the development of the Davisian explanatory framework. If process rates could be identified from current processes and agents then extrapolation and postdiction could be made and the length of Davisian stages calculated. Changing the scale of study, however, permitted collection of empirical data on processes and so the immediate assessment of the operation of environmental systems. A spate of publications in the

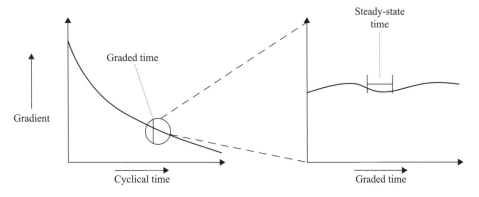

Figure 1.5 Time and its variation with scale, after Schumm and Lichty (1965).

1970s and early 1980s (Chorley and Kennedy, 1971; Trudgill, 1977; Huggett, 1980) all illustrate the acceptance of process studies as a key feature of geomorphology.

This is not to imply, however, that such views were not around before the 1960s and 1970s; Gilbert's work on the Henry Mountains (1877) and dynamic equilibrium was an alternative approach to Davis's historical geomorphology. The development of the physics of sand movement by Bagnold (1941), the association of mathematical modelling and hydrology by Horton (1945), the study of mass movement by Rapp (1960), as well as the research of Strahler (1952, 1966) and Schumm (1956), were all examples of process work in geomorphology. Gregory (1985) assesses the publication of *Fluvial Processes in Geomorphology* (Leopold *et al.*, 1964) as the most influential publication in the 'new' field of process studies. These authors were working within an academic context where process studies were not unknown and where systems analysis was beginning to take hold as an alternative explanatory framework to the cycle of erosion. Their work was important, but it illustrates a trend rather than marks a discontinuity. It could also be argued that Gregory was talking about an academic audience that was relatively small and insular. The textbooks above were aimed at specialist undergraduate courses, presumably taught by academics who agreed with the approach taken. As important, if not more so, are texts that were written at the time for a wider audience. In this context, Cooke's 1984 book *Geomorphological Hazards in Los Angeles: A Study of Slope and Sediment Problems*, as well as Cooke and Doornkamp's 1974 *Geomorphology in Environmental Management*, were early examples of attempts to apply concepts and practices in physical geography to an audience beyond the academic. Indeed, the applied work of Brunsden, Cooke, Doornkamp and Goudie in geomorphology within the UK in the 1970s and 1980s could be viewed as an attempt to develop an applied niche for geomorphological analysis within civil engineering projects and environmental management in general. By playing to this wider audience, it could be argued that these publications are of more significance in promoting the working practices of physical geography than are the textbooks above. Similar arguments could be made for US counterparts in applied work such as Graf and Schumm just to name two. In this manner these workers were applying practical geographic knowledge in much the same way as their nineteenth-century counterparts.

In addition, the significance of graduates pursuing a non-academic career has rarely been considered in subject histories. Graduates who take up posts in consultancy or in management are in positions to greatly influence the type of physical geography applied to 'real'-world problems. The trajectory of such individuals and their use of the academic tools they learned has not really been discussed within the literature of physical geography. This may partly be due to the perception of physical geography as a non-vocational subject. The potential impact of these key figures in promoting and implementing physical geography practices could have aided, via feedback to academics, a more policy-orientated approach to the subject with all the baggage of moral and ethical dilemmas that this entails.

The development of a process orientation to physical geography meant that the definition of the discipline itself began to change. As each subdiscipline of physical geography became acquainted with the new focus, Gregory argues through specific textbooks (*Desert Geomorphology* by Cooke and Warren, 1973; Embleton and King's, 1975a,

1975b *Periglacial and Glacial Landforms* and Carson and Kirkby's, 1972 *Hillslope Forms,* for example) that the need to reach into other disciplines for the necessary theoretical, field and increasingly mathematical techniques for analysis, increased. This meant that the core of study of physical geography became harder to define. The quest for an integrative explanatory framework for physical geography became somewhat superseded by a search for disciplinary homes within their own, existing frameworks. Whereas under the cycle of erosion, when the reconstruction of the environment was the core of study, once process became the focus of study, each subdiscipline became concerned with process agents. Studying these could involve, for example, the analysis of mathematical models of slopes, the detailed physics of wind motion, and the habits of limpets in weathering rocks. The common focus of study disappeared in the melee of process agents. Although it could be argued that the spatial variation of these agents gave each study its geographical flavour, it could equally be argued that the geographer only added an additional viewpoint to an existing field. This problem of discipline definition is a matter that physical geography still has not been able to answer.

Important to the development of a process basis for physical geography was the development of techniques for quantifying landforms and the landscape. The trend towards the increasingly complex representations of the physical environment within each subdiscipline further added to the separation of the subject. Within climatology and ecology, this trend towards a numerical description of processes had long been underway. Climatology had established its data gathering credentials in the nineteenth century and had encountered, and dealt with, albeit in its own way, problems of representation and scale long before geomorphologists began to grapple with these issues. Likewise, ecology had developed techniques to identify and even simply count the entities of interest at a range of scales by the time geomorphologists got around to considering the problem of what to measure. Quantification of the landscape, or rather components of the landscape, was essential to enable an assessment of the contribution of specific processes to the operation of the landscape system.

Despite the great advances made in technology, data collection and a re-orientation towards process studies, physical geography in this period still retains the roots of its past. Despite the new theoretical framework of plate tectonics, physical geography, or rather geomorphology, still retained a focus on landscape development, but at a more human scale. This promoted studies of landforms, and land-forming processes within such features as landslides, which were more generally relevant to environmental management. Despite the greater volume of information, the push for relevancy and the focus on process studies, the basic tenets of physical geography remained the same. The search for universality, the emphasis on the empirical and a concern with change, albeit as equilibrium and process-response rather than as Davisian stage, remained the focus of work in physical geography. The perennial problem of scale remained a sticking point for the integration of processes found at different scales and acted as a brake on a purely reductionist view of the scientific endeavour in physical geography.

A further trend that has been identified by several authors (e.g. Frodeman, 1996; Rhoads and Thorn, 1996c; Lane and Richards, 1997; Baker, 1999; Lane, 2001) is an increasing recognition of the importance of philosophy for the practice of physical geography. Although this may not be a surprise given the purpose of this book, the manner in which this concern is reflected in the work of physical geographers should not be expected to be consistent. The new concern with philosophy does not reflect a new search for a universal framework for explanation, but rather a questioning of the ability of physical geographers to reach such a consensus. One outcome of this new concern has been the recognition of the importance of studying small areas in great detail, for example the significance of small-scale studies as noted by Lane (2001) in relation to fluvial geomorphology. Focusing at this scale of local detail, the universal and contingent become intertwined. Explanation and prediction become more difficult but also more interesting as the application of a general solution is not possible without the 'local' also entering into the equation.

A further trend in physical geography has been the adoption of chaos and complexity as part of the explanatory arsenal.

Chaos, complexity and Earth Systems Science (ESS)

Recent developments in physical geography have seen the introduction of chaos and complexity as explanatory frameworks. These frameworks will be examined in detail in Chapters 7 and 8 but their increasing influence can be seen from the growing use of terms such as 'non-linearities', 'self-organisation', 'emergence' and 'emergent behaviour' (Murray *et al.*, 2009; Church, 2010; Phillips, 2011). These terms throw into sharper focus a continuing debate within physical geography: the ability to understand a complex whole from the study of its parts. Chaos and complexity highlight that even simple rules or laws can produce complicated and unpredictable behaviour, so the detailed study of parts of a system may not provide the information required to understand and predict system behaviour. Likewise, once the parts are put together, behaviour may emerge from the whole that was completely unpredictable from the individual components. The system itself becomes the appropriate level or unit of study and research has to focus on understanding that entity at that level rather than reducing research to an ever-more detailed study of the minutest system component.

Another recent trend is the development of Earth System Science. This, according to Pitman (2005, p. 138), is 'the study of the earth as a single, integrated physical and social system', through the study of which solutions can be provided to major world issues that the reductionist approaches of the past have hitherto failed to resolve. The global and integrative thrust of this science has partially been the result of the desire to provide an overarching framework within which to ensure global sustainability by developing a 'new' science that focuses on the interactions between global social and ecological processes (Kates *et al.*, 2001). Likewise, Komiyama and Takeuchi (2006) suggest that the trend toward a global science is partly based on the need for a science of sustainability that recognises the link between economy and science but which is free from political basis that currently stifles debates to resolve sustainability (International Council for Science, 2002; Clark and Dickson, 2003; Swart *et al.*, 2004). Despite the ambition of a global-scale science, the role of physical geography, when considered, is rather less noble. Pitman (2005) suggests that physical geography can usefully aid Earth System Science by improving the understanding of the biophysical processes and, through this improved understanding, provide climatic modellers with information on the spatial variation of key parameters. This future view of the role of physical geography, as a handmaiden to climatic modellers, is not necessarily one that appeals!

Komiyama and Takeuchi (2006) identify two key obstacles to an earth science – the complexity and interconnectivity of the problems and the fragmentation or specialisation of research. They suggest that overcoming these obstacles requires knowledge structuring – the clarification of relationships between problems and the organisation and mobilisation of specialist fields to answer these refined problems. Although they do not state how this clarification of problems and the integration of diverse specialisms can be achieved, they do propose that there is no single solution to these global problems at specific locations, so solutions have to be sensitive to the needs of place – a very geographical concept.

Clifford and Richards (2005) identify the same trend as Pitman (2005) towards Earth System Science (ESS) but they suggest that despite the explicit holistic goals of the project, the reality of ESS harks back to reductionist approaches. Specifically, ESS attempts to research the particularities of ecological relationships and then pass them into global models of change. They suggest that this approach removes the possibility of solution place-sensitivity and replaces it with a manageable modelling-based and technocentric approach to global solutions which, they argue, will be impractical.

> Perhaps one of the most disconcerting aspects of ESS which issues like this embody is its apparently homogenizing, normative and nomothetic project, possibly as an unconscious attempt to 'make' a more complex world more manageable. In its appeal to a combination of abstract generalization and complex modeling, and to a technocentric fusion of the physical and social worlds, ESS may yet prove to be a reinvention of scientific privilege and practice, with both a language and mode of expression which are restricted rather than general, and where the claims to apprehend the real and complex are, once more, by the few on behalf of the many.
>
> (Clifford and Richards, 2005, p. 382)

A less ambitious, in the sense of being more local than global in its objectives, move towards integrating physical geography with social systems can be found in the literature on ecosystem services. Potschin and Haines-Young (2011) suggest that ecosystem services offer an anthropocentric and utilitarian paradigm for the study of conservation through ecology and economics. The ecosystem services approach (e.g. Millennium Ecosystem Assessment, 2005; Turner and Daily, 2008; Lamarque *et al.*, 2011) is a means of developing integrated solutions to the problem of researching ecosystem degradation. The concept of a service cascade (Haines-Young and Potschin, 2010; De Groot *et al.*, 2010; Salles, 2011) tries to convey the idea that there is a production chain linking ecological and biophysical structures and human well-being. From this framework, key questions can be identified such as the need to assess and monitor critical levels of natural capital required to ensure the continuation of ecosystem services, the importance of 'intermediate' services to the delivery of the 'final' service, and the significance of flows of natural capital. Within this approach the need to value natural capital and its contribution to human well-being is a central concern.

Summary

Ideas in physical geography differ over time and space. Development of understanding about reality can follow one of three models: progressive, paradigms or research programmes. The first model contends that a progressively better and truer representation of the world is achieved by building upon the successes of past researchers. Knowledge is cumulative and progresses towards an absolute understanding of reality as it really is. Paradigms focus a group of researchers around a specific set of ideas about reality. Paradigm choice is as much about social pressures as about logical, objective choice. There can never be certainty that the understanding derived from one paradigm provides a true representation of reality. Confidence that specific paradigms do provide a true representation of reality may be high, but they can never be absolute. Research programmes highlight that researchers do cluster around specific ideas, but their choice of these ideas is based on logical and objective grounds of theory success. The status of these programmes as true representations of reality is problematic and unlikely to be absolute.

Although it is not possible to write *the* definitive history of physical geography, it is possible to pick out trends that seem to permeate much of the history of the study of the physical environment. Three key themes are identified: the search for universality in explanation, the study of stability and change in the physical environment, and the primacy of empirical information to study the environment. The three themes are related and have evolved together, each affecting and encouraging changes in the others. From these themes have developed the notions of a uniformitarian-based approach to processes in the physical environment. Likewise, the rise of Darwinian evolution and evolutionary interpretations of change in the physical environment have developed from the study of contemporary forms and processes. Central to supporting these ideas was the collection and classification of empirical data about natural phenomena. Into the late twentieth century, these themes survived, but the questions asked altered. Process-based studies and a concern with complexity echo the themes of universality and change, but with increasingly complicated empirical data to analyse the physical environment. Recent developments in chaos and complexity have added a potentially new and novel explanatory framework for physical geography. Likewise, the development of Earth System Science and an increasing focus on environmental problems such as climate change, sustainability and hazards has provided the opportunity for physical geographers to illustrate the versatility of the science.

Chapter 2

The nature of reality

What is reality?

Before analysing how physical geographers understand and measure the world, it is important to discuss the nature of the reality physical geographers believe they study. This may seem like an odd place to start, as it seems clear that the physical environment is a real and solid thing that, although complex, is open to study by our senses and the various instruments we have designed to enhance our senses. Such an initial assumption immediately sets up the concept of an objective, real reality of which physical geographers can have knowledge. Indeed, improvements in methods and theories to explain this reality could eventually result in a complete understanding of the physical environment. With this view it is easy to fall into the trap of believing that what is sensed and measured is real, that there is a direct and absolute correspondence between what we think exists and what really exists. All our models of reality capture at least part of a true representation of reality. Refining our measurements or theories, or both, results in a 'truer' representation of reality. In this view of reality, scientific progress means increasing our knowledge towards a true representation of reality, an absolute understanding of it.

The correspondence view of reality outlined above is difficult to sustain in the light of wholesale changes in concepts in areas such as geology, geomorphology and ecology. The introduction of concepts such as plate tectonics, complexity theory and non-selective evolution all imply that previous views of reality were wrong at worst and incomplete at best. How can we be sure that these 'new' concepts are true representations of reality and that the old ones were wrong? The old concepts were thought to be true representations of reality, so their replacement implies that there is something wrong with the correspondence view.

A more appropriate view of reality may be that our current theories provide us with a coherent and non-contradictory representation of reality. In this coherence view, theories seem to work, they seem to explain what is sensed and measured, but there is no certainty that they are true representations of reality; theories can be subject to change. This means that theories tend to be coherent with our current state of sensing the physical environment, and with our current ideas about how the physical environment works. An important question, however, is how do you select which theories are the most coherent? What criteria do you use and who has the final say? The arguments become very similar to those for and against paradigms and research programmes. The scientific community seems to develop and agree theories, but this is not necessarily a guide to how true those theories are. Despite the seemingly conditional nature of theories, there is an implicit assumption that somehow contemporary theories are truer representations of reality than older theories. Science improves and progresses, our theories begin to mirror reality in a clearer fashion. Even if we recognise that we have not got an absolute understanding of reality, we feel we have a better grasp of it than we did in the past. Unfortunately, there is nothing in a coherence view of reality to support this assertion. All we can say is that our current theories seem to cohere with what we sense; we still do not have any absolute basis for believing our theories are better, or are a closer representation of reality, than theories in the past.

Being extremely pessimistic we could just plump for a more utilitarian definition of reality. Our theories make predictions that happen, our theories produce useful results so they seem to be useful: the pragmatic utility view of reality. This view tends to be seen as the least acceptable by scientists. It makes no claims about any movement towards an ultimate knowledge of reality. It makes no claims that our theories are more coherent than theories of the past, nor that coherence with current beliefs is the most appropriate way to assess the worth of a theory. Instead, usefulness replaces the quest for truth as the basis for theory selection and representation of reality. Taking a pragmatic view is not, however, necessarily a denial of a consistent working method for scientific investigation. Rescher (2001), for example, outlines functionalistic pragmatism as a possible approach to scientific investigation. Theory is central to this view of scientific investigation. The success of a theory is judged on how successful it is in its practical implementation. Success, of an instrument, or methodology, or procedure, lies in its successful application. Functionalistic pragmatism is not concerned with the truth of reality, it is concerned with how truths about reality are thought to be identified, with the processes of scientific endorsement. It is concerned with how and why 'truths' are validated. Success is defined by how well practices provide answers to our particular goals. Importantly, however, Rescher makes the point that although goals and purposes are important and can be matters of taste, evaluation is not. Evaluation requires a rationale, it requires rules and procedures that permit comparison of 'truths' between individuals.

The above views do share a common foundation by accepting that there is an external reality that is capable of study. We may not be able to get a real or true representation of that reality, but there is definitely something out there to study and, more specifically, to interact with. In addition, there is an implicit assumption that the reality we study has some causal basis. There is a reason, which is discoverable, for why something happens. Events in the physical environment are determined by other events and mechanisms; it is a deterministic universe. Such a view stands in distinct contrast to some of the philosophical stances of human geography, such as humanistic views of reality. The idealists' view of reality as constructed in our own heads, the phenomenological view of reality as about deriving essences, and the essentialist pessimistic view of reality as angst and woe do not assume that there is something external to humans. The assumptions of being able to understand reality without reference to some external and real environment are difficult to find counterparts for in physical geography. Although some have suggested that a quantum mechanics-based view of physical geography implies an idealist view of reality (Harrison and Dunham, 1998), reality does not exist until we measure it or interact with it; there have been few, if any, attempts to implement such views. Assuming an external reality does not necessarily mean that scientists have to assume they will eventually understand that reality in full. Likewise, this assumption does not mean that scientists need assume that their representations are of reality as it really is. What the assumption of an external reality does provide is a basis for intersubjective communication – a basis for discussion within a common framework. One scientist may believe they have identified a process or an entity. Their analysis will probably be based on commonly agreed practices and involve criteria for identifying and manipulating the process or entity. This means that other scientists have a basis for recreating and redefining the process or entity identified by any single scientist. Hall (2004), for example, used data from short-term measurements of temperature variations in a stone block to identify the release of latent heat when water freezes in the stone. He used this information to suggest that standard views of freeze-thaw were inappropriate as they overestimated the frequency of this weathering process. The proposition is supported with evidence from sensed data that is available to other scientists to reinterpret, Hall noting that one reviewer suggested that instrument sensitivity or reliability might explain the slight change in temperature measured. Similarly, Hall describes the monitoring set-up in sufficient detail for other scientists to repeat his whole monitoring procedures to assess if they get similar results. Hall's assertion is, in other words, assessable by other scientists because they all assume what he is observing is some aspect of an independent reality that is open to interpretation, both in how it is sensed and in the replication of observations. Without assuming that there is a common reality, such refinement or even redefinition of phenomena would not be possible.

Views of different philosophies

Despite the assumption of an external reality, there is no single philosophy covering all of physical geography. This problem is made more acute by some attempts to graft various philosophies onto what physical geographers do. Reviewing, with the benefit of hindsight, 'old' masters or even current practitioners of physical geography as really practising a contemporary philosophy without knowing it, is fraught with the same problems as developing Whiggish histories of the subject. It is important, however, to be clear about the different philosophical standpoints that have been advocated in physical geography and to be aware of their similarities and differences. Central to identifying the nature of the four philosophical standpoints considered here, logical positivism, critical rationalism, critical realism and pragmatic realism, are the ontology and epistemology of each. Ontology refers to what each philosophy believes is real. Raper (2000) identifies ontology as being concerned with concepts of identity – the way in which reality is identified and ordered. Epistemology refers to how each philosophy believes we can know that what we think is real is real. It is the study of how knowledge can be acquired, how it can be validated. Together these provide the basis for a detailed consideration of each philosophy.

Logical positivism is usually held up as a relatively naïve philosophy with little to recommend it in contemporary physical geography. Its ontology is that what is real are the laws and entities that interact in reality. Reality is as it is because the laws we believe exist really do exist and these underlie the operation of all entities that also really exist. We can know about this reality by observing or sensing the interactions of entities through the operation of laws. This is achieved by observing and, where possible, measuring changes or the lack of change in entities under the action of specific laws. From such an analysis we can derive the structure and nature of reality as it really is. Logical positivism has its roots in the Vienna circle of philosophies in the 1920s, although as a view of how science works it goes back further, depending on how far you want to push the argument that early scientists worked with a philosophy as coherent and consistent as outlined by the Vienna circle. Central to the logical positivist is the idea that you can identify, observe and measure real entities and their interactions. From these observations it is possible to derive laws of behaviour that can be proven to be real by further observation.

Critical rationalism, developed mainly by Karl Popper (1968) as an alternative to logical positivism, has the same basic ontology and epistemology. Critical rationalists believe that there are entities that interact in a regular manner that are capable of observation. The key difference between the two philosophies lies in the belief of the possibility of obtaining truths about reality. Logical positivists believe that it is possible to assess a theory and determine if it is a true reflection of reality: 'we believe something because we can prove it'. Popper's critical rationalism denies that it is ever possible to be certain that a theory really does reflect reality. All we can do for certain, according to Popper, is to show that a theory does not reflect reality. Theories can be disproved, falsified, but they can never be proved. Falsification of a theory becomes the hallmark of the difference between science and non-science. This view does, however, tie Popper to a continual conditional view of reality. There can never be absolute certainty that our theories match reality. There is no gold standard of truth to which to compare our ideas. This could be seen as a problem as it implies that even cherished and seemingly correct theories and their outcomes, such as gravitation and the rising of the sun, are not absolute certainties. Popper suggested that although theories could not be proven, there could be increasing certainty of agreement with reality, increased verisimilitude or truth-value, but never really stated how this worked in practice.

There has been a conscious attempt within many areas of physical geography to adopt a Popperian approach to research. This stems from the belief that this explanatory framework offers a more robust basis for inference; hypotheses that have survived attempts at falsification increase in stature and the science moves forward. In palaeoecology, for example, and arguably in other subdisciplines of physical geography (Shennan, 1995), a long period of curiosity-motivated science, or research specifically conducted to address 'areas of ignorance', such as the cause of the mid-Holocene elm decline, or the nature and timing of Late Quaternary vegetation changes (Oldfield, 1993), has provided the necessary foundation and opportunity for hypotheses to be generated, tested and modified accordingly. Thus, the nature of the questions that scientists want to ask has evolved as knowledge of their subdisciplines has grown. The attempt at the falsification of competing hypotheses provides a clear impetus for scientific investigation, and defines which research strategies should be adopted. Research conducted within this framework is seen as more purposeful, as it discourages research programmes to suit the comfort of the researcher (e.g. in applying established techniques rather than developing new ones, or in working within the confines of their narrow specialism, rather than embarking on interdisciplinary research when required). Falsification as a scientific approach in physical geography possesses a certain *gravitas* – and many workers see it as the gold standard.

At the heart of both the logical positivists' and critical rationalists' views of reality are the use of induction and deduction in developing scientific explanation. Inductive reasoning believes that the 'facts' can speak for themselves. This could be seen as the 'vacuum cleaner' approach to science. Collect as many facts as you can and, from this mire of data, theories will spontaneously form and emerge full-grown into the world. Along a similar vein, inductive reasoning assumes that the truth of a statement is proportional to how often it has been shown to be correct. Statements become true by collecting data that demonstrate that they are true. This form of reasoning puts a premium on data collection as the source and basis of theory and truth. Unfortunately, it is a logically flawed approach. Rarely, if ever, are data collected without a reason, without some guiding concept, that determines what type of information should be collected and how. Likewise, if a single counter-example to a theory is collected this destroys the logic of inductive reasoning. A single data point that does not fit the theory developed reduces the truth value of that theory, breaking its supposed explanation of other data.

Deductive reasoning is based on the development of a logically, internally consistent argument. This approach has been called syllogistic reasoning: a conclusion follows logically and inexplicably from two premises. In its guise as the covering law model, the two premises are a set of initial conditions and a covering law. Applying the covering law to the initial conditions, the conclusion is an inevitable outcome. This form of reasoning provides consistent and logical statements. The only problem is that these statements need not bear any relationship to reality, as outlined in Table 2.1. Popper, however, used this structure of reasoning in his philosophy of critical rationalism. The derived statement could be viewed as a prediction of how reality should be if the covering law operates. It provides a statement that could be compared to reality to see whether the covering law works under those initial conditions. Popper viewed the derived statement as a hypothesis capable of being tested. Unlike the logical positivists, however, Popper was intent not to prove the statement true, but to devise ways of disproving or falsifying it, as discussed above. Falsification became the key criteria for judging whether a theory was scientific or not. If no statement could be derived that was testable, then the theory was non-scientific. A standard comment at this point is that this meant that Einstein's theory of relativity was defined as non-science until 1918 when its predictions could be tested by a solar eclipse. This is not quite the case, as the theory of relativity was capable of falsification. There was an experimental situation that could be envisaged with the existing technology in which the theory could be tested, in which it could be capable of falsification. The specific circumstances required to undertake the crucial experiment had not been available until the solar eclipse of 1918. This makes the rigid demarcation between science and non-science much hazier, as capable of falsification is a more difficult idea to pin down than actual tests of falsification.

Table 2.1 *Deductive reasoning.*

A

Initial conditions:	Gerald is a budgie
Covering law:	All budgies are killers
Conclusion:	Gerald is a killer

B

Initial conditions:	Rocks in coastal environments are more subject to cycles of wetting and drying as the tide goes in and out than rocks in inland locations
Covering law:	Alteration of rocks by wetting and drying increases as the number of cycles increases
Conclusion:	Rocks in coastal environments are more severely altered by wetting and drying cycles than are rocks in inland locations

Argument *A* is logically consistent, but unless you believe budgies are killers, is totally ridiculous. Statement *B* is logically consistent, but also unlikely. The statement refers to only a single process that is more dominant in its operation at the coast. The problem is that this process co-varies with other processes such as salt weathering. Despite the logic of the statement, the researcher requires an a priori knowledge of reality to assess or test the 'truth' of logically valid statements.

Falsification also implied that there is never any certainty about the theory you are trying to falsify. Popper's ideal was to test competing theories until one could not be falsified. This was Popper's method of multiple working hypotheses. Different theories generate different covering laws that when applied to the same initial conditions should result in different, testable hypotheses. Each hypothesis should be capable of being falsified by critical testing. Ideally, you should be able to eliminate every hypothesis bar one by such critical testing. At this point you accept the hypothesis that has survived falsification, and its associated theory, as provisionally correct. The theory was never accepted as true, never accepted as proven, just not capable of falsification at this point in time. This means that any theory is unprovable. Achieving absolute truth, getting to reality as it is, is not possible in Popper's scheme. Popper suggested that although no theory could be proven, as a theory withstood more tests to falsify it, confidence in that theory grew. The truth value of a theory increased the more testing it survived, but it never became true. This rather unsatisfactory end point for a theory implied that key scientific ideas such as the law of gravity were not necessarily true, just unfalsified. Their stability as cornerstones of scientific thought came from the confidence derived from their continued non-falsification, and the lapse in interest in trying to falsify them resulted from this confidence.

There is an understandable tendency for scientists to hold in high regard hypotheses that have survived falsification. However, failure to falsify a hypothesis may arise entirely as a result of asking the wrong types of question, or in applying a research design that is flawed. There is then a danger that weak hypotheses may attain a high status; its high status may dissuade further attempts at falsification, and a hypothesis then becomes entrenched, perhaps even leading to a paradigm shift. Elner and Vadas (1990) contend that this has happened in ecology. Specifically, they believe that the establishment of the paradigm that the American lobster is a keystone species emerged without sufficient investigation or appropriate testing. The lobster keystone predator hypothesis emerged following the population explosion in sea urchins along the Atlantic coast of Nova Scotia during the early 1970s, which was concomitant with a decline in the commercial landing of the American lobster. However, after reviewing a wide range of literature detailing the results of scientific investigations into this issue, Elner and Vadas (1990) identified several key areas of concern. For instance, results that contradicted the keystone-predator paradigm for the American lobster tended to be explained away; the results were ignored because they were argued to have no direct bearing on the 'real-world' situation. Additionally, reference to the American lobster as an example of a keystone predator entered textbooks, reviews, and popular articles – it pervaded popular thinking – and thus became further entrenched. Important processes and relationships that underpinned the hypothesis were based on evidence that they deem insufficient (e.g. the perceived intensity of fish predation of sea urchins). Such processes and relationships became 'truths' in the scientific literature following repetition over several publications, thus obscuring weaknesses in the original evidence. Elner and Vadas (1990) conclude that the scientific community must exercise caution against allowing paradigms to become established without adequate scrutiny.

In a similar vein, scientists must also ensure that potentially strong hypotheses are not dismissed prematurely – contradictory results must be shown to be valid before being accepted. Attempts to falsify the Early Anthropogenic hypothesis of Ruddiman (2003) represents a case in point. The hypothesis proposes that human activity has resulted in a reversal in the otherwise natural decreasing trend in atmospheric carbon dioxide (CO_2) and methane (CH_4). Specifically, an increase in CO_2 from $c.8,000$ years ago, and in CH_4 from $c.5,000$ years ago, as a result of forest clearance and rice irrigation respectively, led to a small amount of anthropogenic warming sufficient to reduce the effects of natural Late Holocene cooling, and thus prevent the onset of a new glaciation within the last several thousand years. This hypothesis is based on the interpretation of a wide range of geological and archaeological evidence. Others contend that the rise in atmospheric CO_2 and CH_4 can be explained by natural processes (e.g. Indermuhle et al., 1999), and suggest that the scale of anthropogenic activity was insufficient to lead to the observed rise in atmospheric CH_4 and CO_2 (e.g. Joos et al., 2004). At the heart of the issue is whether the rise in mid- to late-Holocene CH_4 and CO_2 was natural or human-induced. The protagonists of each view have attempted to falsify the competing hypothesis proposed to account for the rising atmospheric CH_4 and CO_2 concentrations. For example, several papers claim to have falsified the anthropogenic hypothesis as a cause of increasing Late Holocene atmospheric CO_2 concentrations (e.g. Joos et al., 2004; Elsig et al., 2009; Stocker et al., 2010). Yet Ruddiman et al. (2011) maintain that claims of falsification were based on invalid evidence, and in fact, neither an anthropogenic cause nor a natural cause can be falsified based on the current data available. Ruddiman et al. (2011) are unable to falsify an anthropogenic cause for the observed rise in late Holocene CH_4 concentrations. In particular, they point to the downward trend in atmospheric CH_4 observed in previous interglacials as evidence that the upward trend in observed atmospheric CH_4 concentrations during the late Holocene is anomalous, and thus unnatural. This brief consideration of the adoption of falsification as a working method in physical geography demonstrates both its potential to generate new research programmes in instances where hypotheses cannot be falsified, but, perhaps as a victim of its own success, its potential fallibility – in the sense that research conducted using this approach, including research that may lead to the apparent successful falsification of a hypothesis, may be automatically held in high esteem by association, even though in reality this research may perhaps not be sufficiently rigorous.

Case Study

Critical rationalism: an example from environmental reconstruction

Battarbee *et al.* (1985) discuss the possible causes of recent (post *c*.1850) lake acidification in a specific region of southern Scotland. They identified four possible causes of recent lake acidification for this part of southern Scotland: long-term natural acidification, afforestation, heathland regeneration, and acid (anthropogenic) precipitation. For each possible cause a plausible scenario could be developed linking effect to cause. Natural lake acidification over several thousand years as a result of lake ageing had already been identified in other temperate environments. In this scenario, however, acidification of lakes is likely to have increased little since 1800. Afforestation would alter the soil characteristics and increase the acidity of runoff to lakes. Heathland regeneration, specifically an increase in *Calluna vulgaris* in Galloway in the wake of a decline in grazing, would again alter soil characteristics and so increase the acidity of the water reaching the lakes. Finally, acid rain from industrial and other activities would increase lake acidity by direct input from precipitation.

These four causes could be tested by the analysis of lake sediment deposits from Loch Enoch by looking at changes in diatom populations, pollen composition and heavy metal content. Diatoms are a group of microscopic algae with high preservation potential, owing to their siliceous (opaline) cell walls. They occupy benthic and planktonic lake habitats and are sensitive to changes in lake pH. Therefore, analysis of the fossil diatom assemblages preserved in the sediment deposits of Loch Enoch permits an estimation of changes in lake pH over time. Pollen grains, at least those that enter the lake and are preserved in the sediment, indicate the nature of the plant community within the lake catchment. Heavy metals are not found in significant quantities within the lake catchment. Any heavy metals within the lake sediments are therefore likely to have been transported into the catchment, probably via atmospheric transport and subsequent wet and dry deposition. Their presence points to pollution sources external to the lake catchment. Each of the above causes of lake acidification would be expected to alter the nature of either the diatom or pollen communities, or the amount of heavy metals. This means that each cause can be rephrased as a simple question, a hypothesis, to which there is a simple yes or no answer – a falsifiable question.

The four causes need to be rephrased as hypotheses and the criteria for falsification clearly identified. Long-term lake acidification can be restated as:

> Do diatom communities indicate that acidification has been either increasing steadily or has not increased in the last 200 years?

This question takes the data from diatom analysis as the basis for falsification. The question defines a specific time period to which the assessment should be applied and identifies the two conditions that would mean this cause could not be falsified. Diatom analysis suggests that acidification began in 1840, so the above question can be answered as no to both parts.

Afforestation as a cause can be rephrased as:

> Does the recent (last 50 years or so) afforestation increase the acidity of lakes in this region as indicated by diatom communities and by evidence from other lakes in the region?

This hypothesis is assessed using not just the sediment core data but also using information from other studies. The investigators reject this hypothesis by referring to the acidification of non-afforested catchments, such as Loch Enoch, as well as the evidence of acidification of other lakes beginning in the late nineteenth and early twentieth centuries, before afforestation of the lake catchments.

The land-use hypothesis is assessed using the pollen in the sediment core in combination with the evidence of the timing of acidification provided by the diatom community data. The hypothesis can be rephrased as:

> Does an increase in heathland area, as determined by pollen analysis, result in an increase in the level of acidification of the lake as indicated by the diatom communities?

The pollen data suggest that the plant communities have been relatively stable over the last 200–300 years and, if anything, *Calluna vulgaris* has declined in the recent past, not increased.

Finally, the investigators test anthropogenic acid rain as the cause of lake acidification. The cause can be rephrased as:

> Do heavy metal concentrations in the sediment core increase as lake acidification increases, as indicated by the diatom communities?

The question requires the identification of fluxes of heavy metals, specifically zinc, copper and lead, all metals that would not be expected to have high natural concentration in this type of catchment. The only source for these metals that the investigators entertain is from industrial sources. Background concentrations of these metals from the catchment show no abnormalities. The increase in the concentration of these heavy metals in the sediment core co-varies with the acidification of the lake as indicated by the diatom communities. This means that the question can be answered as yes, and so the hypothesis is not falsified. The conclusion of the paper is worth quoting in full however:

> We cannot prove that an increase in acid deposition was responsible for the acidification of Loch Enoch and similar lakes in Galloway, but we have shown that alternative hypotheses, as presently formulated, are inadequate.
>
> (Battarbee *et al.*, 1985, p. 352)

Nothing has actually been proven as true. The inability to falsify the last hypothesis is viewed only as indicative of what is likely to be the 'real' cause. There is the potential for other evidence and, as importantly, other formulations of hypotheses that could be assessed and falsified. Acceptance of acid precipitation as the cause of lake acidification is only provisional in the absence of other candidates. The conclusion is scientifically accurate in Popperian terms. It does not extend its results beyond the Galloway region, which is an area of similar geology, ecology and remoteness as the study site.

Recently, critical realism (Bhaskar, 1978, 1989; Collier, 1994) has been put forward as another possible philosophical framework for physical geography. Although there are many flavours of critical realism, there are certain important differences in their ontology and epistemology compared to critical rationalism. Critical realism views reality as both stratified and differentiated. Reality is made up of entities that interact according to underlying laws related to underlying structures; it is differentiated. This differentiated reality is also stratified. Entities at the level of physical and chemical particles are subject to laws at that level. Organisms, being composed of molecules, are subject to the same laws, but in addition their interaction produces new laws found only at that level of reality. These new laws are constrained by the laws at lower levels. No organism can behave in a manner that would break or contradict physical and chemical laws. Within these constraints, new relationships emerge that, although inherent within the lower levels, are not predictable from them. This means that laws at one level or strata of reality cannot be reduced to laws of another level. Evolution, for example, is not reducible to purely genetics, although genetics plays a role. Evolution occurs at the level of the organism and as such it is the relations at this level that determine the nature and regularities of evolution. This also means that causation and explanation need to be appropriate to the level of the entities of interest.

A differentiated and stratified reality means that the level of our observations is not the only level at which explanation can or should occur. All reality is encompassed at the level of mechanisms and structures. These structures and mechanisms provide the range of possible relations that could produce events. Below this level is the level of the actual. These are the events actually produced by the interaction of structures and mechanisms. Below this level is the level of experience. These are the events that we can identify and sense. As you go down from the level of structures to experience, there is a decrease in the events available for study. At the level of structures, there are a vast number of possible events that could happen. Structures do not exist in isolation. Interactions between structures eliminate some events, whilst permitting and even generating others. These interactions restrict the types of behaviour possible and so reduce the potential events that structures could generate individually. This contingent and unpredictable juxtaposition of structures produces events that actually happen. These events further define and constrain the future interactions of structures and so the actual can influence the development of the structural level. The level of the actual is not, however, the level that is studied. The level of experience is a much smaller range of events that we can identify and sense. Positivism and critical rationalism tend to reduce reality and explanation to this final, empirical level. At this empirical or experiential level, the pure operation of structures is not likely to be clear. The power of individual, underlying structures to generate events has been influenced, potentially even nullified, by the interactions with other structures. All that is likely to be observed is a weak signal of the potential behaviour generated by an underlying structure.

Pragmatic realism and semiotic approaches to physical geography have recently been suggested as an appropriate framework for understanding in physical geography. Baker (1996a, b, 1999) looks back to the founders of geomorphology in North America and the practically-minded philosophy of Charles Pierce (1957), semiotics, as a basis for developing a pragmaticist approach. Pragmatic realism is not a simplistic philosophy about usefulness being the measure of worth – this is the pragmatic utility view of reality. Ontologically pragmatic realism believes that an external reality exists with entities that interact with each other. Epistemologically, as researchers, however, we can never be certain we have divided up reality as it is. All we know is that we have divided it up in a way that seems reasonable and consistent for our purposes – it is pragmatic in this sense. There may, however, be a number of ways in which reality could be divided and studied that are consistent and pragmatic; there could be a number of ways of world making. Pragmatic realists accept that this is the case and that their models of reality are conditional and subject to debate and change. Having accepted this, pragmatic realists still believe that reality is underlaid by structures that produce law-like behaviour. We can theorise about these structures and laws, but as with entities we can never be certain that we have identified reality as it really is. Understanding itself develops by relational thinking. Understanding is only possible because we relate one entity to another; we only understand an entity by its relationship to other entities. These entities are only understood, in turn, by their relations to other entities, and so on.

Baker (1999) believed that physical geography should use the concept of signs to study reality. Baker suggests that, from Pierce, signs involve certain conditions. First, the sign must represent an object in some manner. This representation is based or grounded in the sign itself. The sign provides a quality to the object that exists within the sign. The example Baker uses is of a rock being observed as foliated. The quality of foliation is a quality given to the object, the rock, by the sign. Second, the sign must be interpreted by an interpretant. A sign represents an object only if there is an interpretant to correlate or draw the two together. There is no base to this interpretative process however. The foliated rock, for example, is not just foliated, it is composed of a past history and a current state that are also conveyed in the sign of foliation. The relationship between foliation and the rock texture it expresses is a further sign relationship, a relationship requiring an interpretant. Baker makes the point that this chain of sign interpretants is never-ending. This idea is illustrated in Figure 2.1. The fossil in the figure has a relationship to its object, the past organism that it represents. Baker identifies this as a causal or indexical relationship (Baker, 1999, p. 641). Fossilisation itself produces a kind of interpretant, the fossil within its geological bed. It is sign waiting for realisation. Baker suggests that this sign can be triggered by a trained palaeontologist who can interpret its causal relations. The causal connection between the fossil and its past organism is translated into a mental process of connections in the observer. In turn, these become

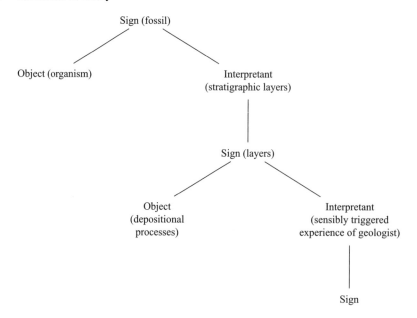

Figure 2.1 Baker's (1999) example of infinite semiosis.

further triggers for products, such as scientific papers, and further inquiry, all interpretable as signs in themselves. The important point of Baker's analysis is that the fossil on its own has no value or worth. It is only when it becomes enmeshed within a network of signs that it can be interpreted and acts as a stimulus for further sign generation. Significantly, this view of reality links the object and subject directly and blurs the distinction between the two. This is a theme that will be developed further in relation to entities (Chapter 3), but it is sufficient to note here that, arguably, this makes a clear objective/subjective division of reality almost impossible within pragmatics. This view of a continual shift between the objective and subjective in the working practice in the study of the physical environment is not new. Van Bemmelen (1962), for example, noted the interrogator of reality, the geologist, needs to constantly move from induction to deduction and back again in pursuing a comparative analysis of observation and evidence.

Reality as a dialogue

The philosophies outlined above provide a snapshot of current thinking in science, but they also reflect a general bias in most discussion of scientific philosophies. The philosophies tend to focus on the 'hard' sciences, and physics in particular, as their model of what science is, and how science should be done. This means that most discussions concerning philosophy may appear to be dealing in the abstract, but often use concepts of direct relevance to the practice of physics. Physical geographers have tended to assume that their approach to reality has to match that of the 'hard' sciences. Problems unique to trying to undertake what is variously described as a field science, an historical science or an environmental science, have been viewed as merely local problems that should not concern the general philosophies used. This has resulted in post hoc justification for research designs and particularly phases of reductionist work in physical geography, in an attempt to ensure that physical geography remains a 'proper' science.

There may be nothing inherently wrong with this view, but if it is taken as the only view of physical geography, the only way to approach this particular and unique scientific endeavour, then it can become highly restrictive. This view tends to perpetuate the idea that there is an objective reality from which we draw real knowledge. Reality is there – we are clever and extract real data from it to find out what reality is really like. This view is one that is difficult to sustain once you start the practice of actually doing physical geography. At this point, another view of how reality is understood may become more appropriate: the view of reality as constructed by a dialogue.

The metaphor that understanding of reality is at heart based on a 'dialogue' between scientist and reality is not a new one, and has been associated with physical geography as an historic science. Demeritt and Dyer (2002) identified this trend in both human and physical geography. They identify different uses of the term 'dialogue' applied to the research strategies of geographers, including literal and metaphorical. The metaphor of 'dialogue' begins to make problematic what appeared to be natural. The metaphor highlights the importance of both the observer and the observed and continually switches between them.

The use of metaphors such as a 'conversation' with nature, or 'questioning' of nature, or of 'reading nature as a book', are common both as teaching methods and as views about how to study nature (e.g. Cloos, 1949; Wright, 1958; van Bemmelen, 1962). Despite the common use of these metaphors, they have been little used to form a basis for trying to devise a philosophy of physical geography and its practice. Although these metaphors can be interpreted as supporting a very reductionist view of reality, implying that there is a single, correct reading of the book of nature for example, they can also be interpreted as implying a less rigid view of reality. Van Bemmelen (1962) noted, for example, that the character of the subject, the researcher, can become as important as that of the object of study. Indeed, he compared the geologist to a doctor highlighting the 'art' of investigation as much as the 'science'.

The dialogue metaphor strikes a particular resonance with the critical realist and pragmatic realist views of reality. The dialogue metaphor highlights the negotiated nature of reality. Reality is not just a thing to be probed and made to give up pre-existing secrets. Reality enters into the dialogue by answering in particular ways and by guiding the types of questions asked. Central to this metaphor is the interpretative nature of enquiry. This does not deny that an independent reality exists, only that it is the engagement with that reality that is the focus of study. Some pragmatic realists have argued that the assumption of a real, external reality is vital for any dialogue to work. Although an absolute knowledge of the external reality can never be achieved, the idea that there exists an external reality means that researchers have an assumed common basis for reference and discussion of their understanding. Each researcher is willing to enter into this dialogue because each recognises the potential fallibility of their own view of reality and so the potential for modification. Reality itself plays a constraining role for this debate amongst researchers, guiding research questions and methods, but not along a predetermined or predefinable path.

The dialogue, rather than the subject of the dialogue, is what we can know about. This puts engagement with reality, or rather physical geography as it is really practiced, at the heart of any philosophical discussion. This also means that the false distinction between 'theoretical' and 'practical' work dissolves, as practice involves both. Similarly, the dialogue metaphor pushes debate away from other polar opposites such as 'realism' and 'idealism', or 'objectivity' and 'subjectivity'. Rather than looking at these as 'either/ or' debates, practice or the 'doing' of physical geography requires the investigator to continually shift from the 'objective' of the entity under study to the 'subjective' of the mental categories and concepts used in order to understand the entity. The continual shifting and informing of practice by the 'real' and the 'mental' means that neither can be understood in isolation. Practice brings to the fore the relational nature of understanding in physical geography and the haziness of seemingly concrete concepts such as entities.

Theory, reality and practice

Defining theory precisely is highly problematic. Some authors provide rigid and highly formalised definitions of theory, whilst others question the value of the term at all. Theory can be viewed as a framework of ideas that guide what we think reality is and how to go about studying it. Abstract ideas such as force and resistance can be thought of as theoretical constructs. These two ideas can be linked together to suggest what should happen in reality. For example, if force is greater than resistance then change should occur. The two ideas form irreducible elements of the theory. They are the axiomatic elements. Their existence is not questioned, it is taken as given within the theory. Likewise, their interactions, usually based upon some set of physical principles or mechanisms, are the axiomatic principles of the theory. The ideas are, however, very abstract and need to be translated or superimposed upon parts of reality. Theory needs to be linked to reality. This is usually achieved by bridging principles, rules that define how abstract ideas can be interpreted as real entities capable of identification and testing. Within science, these bridging principles are essential; they are the means by which theories can be tested and, in a pure Popperian sense, rejected or accepted (see Figure 2.2 for an illustration of the structure of a theory).

It is interesting to note that Haines-Young and Petch (1986) distinguish between theories and myths. Within myths they include Schumm and Lichty's (1965) paper on time, space and causality, Wolman and Miller's (1960) paper on magnitude and frequency as well as Hack's (1960) paper on dynamic equilibrium. For many geomorphologists in particular, these papers would seem to be central to any understanding of the physical environment. Haines-Young and Petch, however, make the point that all these papers express what they term 'truisms'. For Schumm and Lichty, for example, they state that the truism is 'that theories explain variables and that only certain theories have been developed' (p. 115), a problem at the time of the paper but, in their eyes, a trivial issue in retrospect. Wolman and Miller (1960) are merely stating the truism that whatever

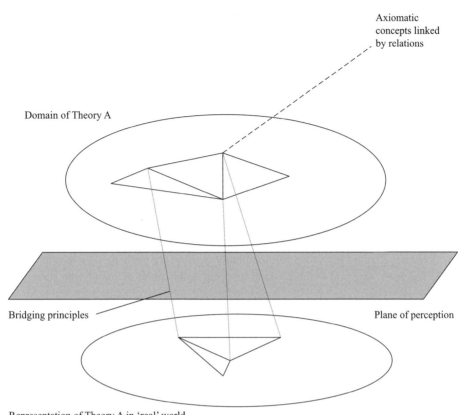

Axiomatic
concepts linked
by relations

Domain of Theory A

Bridging principles

Plane of perception

Representation of Theory A in 'real' world

Figure 2.2 Representation of theory.

landscape feature is observed, it must have been formed by processes that had some magnitude and frequency (p. 139). These papers, as well as Hack's (1960) outline of dynamic equilibrium, state ideas that cannot be disproved. They can be linked to reality in a multitude of ways excluding nothing, and can never be falsified because of this. An inability to focus or narrow down a set of ideas to form a distinct and singular set of bridging principles means that there is always scope for reinterpretation of the ideas in a favourable light, i.e. the ideas are unfalsifiable. Myths are not useless in physical geography however. As Popper indicates, 'a myth may contain important anticipations of scientific theories' (Popper, 1963, p. 38). Moreover, Haines-Young and Petch point out that although these ideas are not theories, as defined within critical rationalism, and so are not sources of any understanding about reality, such myths form the basis of a lot of testable theories in science. They view it as possible to derive testable and coherent theories from the myths. They see it as the task of a scientist to develop such links between myth and theory, to prompt questions based on these myths that are amenable to testing in reality. Where the dividing line is located between the myth and the derived theory is unclear. There may be many ways of undertaking such a translation from myth to theory, and so the problem of an unfalsifiable base remains, despite the presence of a falsifiable product, a theory.

Case Study

Myths and theories

The classic paper on 'time, space and causality' by Schumm and Lichty (1965) is often viewed as a major turning point in thinking about geomorphology. Haines-Young and Petch (1986), however, suggest that the main argument of the paper amounts to a 'truism' or myth rather than a theory. The basic idea is that at different temporal scales, different variables will be independent, dependent or irrelevant for understanding the functioning of the landscape. The temporal scale of analysis you are interested in determines which of the variables you need to study. Undertaking a study of a catchment within graded time, for example, would require an analysis of the relationship between dependent and independent variables. Variables such as time and initial relief are irrelevant for your study as these are unchanging over this timescale. Haines-Young and Petch (1986) characterise this point as merely the statement that geomorphologists have developed different theories to explain landform behaviour at different scales. The variables vary in their status as dependent, independent or irrelevant, because the theories used change. It is not the variables that alter their status, it is the theories that alter to explain a different variable at a specific scale. Tinkler (1985) notes that the key point to Schumm and Lichty's analysis was to highlight that theories actually relate to different things at different scales. Acceptance of one approach, the historical, is not incompatible with another, say that of equilibrium, as each deals with different things at different scales. Each is mediated by different theories that identify different objects of study.

The ideas of Schumm and Lichty are not falsifiable. They are only statements of how research into reality is structured within geomorphology. The idea that theories relate to different variables at different scales does not generate any hypotheses to assess. The idea could, however, provide a basis for directing thinking about reality. If theories relate to different variables at different scales, then an investigator is not bound by an overarching single scale-independent theory. What is an appropriate theory, an appropriate approach to study, could vary with the scale of object being studied. Indeed, the objects themselves could change as you change scale. It is in this sense that the 'myth' of scale put forward by Schumm and Lichty is of use in geomorphology.

Von Englehardt and Zimmerman (1988) suggest that theories are hierarchically ordered systems of hypotheses interlocked by a network of deductive relationships. This highlights the significance of hypotheses to one view of theory and possibly may give myths a role as a higher level structure. They view theory, hypotheses and the empirical levels of reality as interlinked. Using the theory of plate tectonics they construct a hierarchical framework of theories, hypotheses and 'facts'. At the highest level of the theory are the basic hypotheses from which the partial hypotheses are derived. From the partial hypotheses are derived the inductive generalisations and empirical basis of the theory. Each level informs the others. The basic hypotheses are few, but the hypotheses derived from them are many. Likewise, the empirical 'facts' are more in number than the partial hypotheses that they inform. The structure of an increasing number of derived statements is a common one. Both empirical and theoretical 'facts' (by which they mean theoretical concepts such as the atom), can both be used to construct theories. Within physical geography it can be assumed that most theories will contain both and be informed by the success or failure to explain or predict the behaviour of these entities.

Von Englehardt and Zimmerman draw little distinction between theory and hypotheses and regard the two as related. Theories only differ in being further up the hierarchy and being greater in the scope of their application, their significance as explanatory structures, and being more reliable in terms of consistency. According to von Englehardt and Zimmerman (1988), a theory should be both internally consistent, in terms of its logic structure, and externally consistent, in terms of its consistency with the hypotheses derived from other theories. These other theories include the underlying physical and chemical theories that usually make up unquestioned components of theories in physical geography. They do recognise that both theory and hypothesis represent generalisation, abstractions of reality. They are not mirror images of reality, they describe and explain only the empirical and hypothetical components of reality and so are only indirectly reflecting reality as it is.

Theories can be identified at a number of levels in any study. A theory about force and resistance within fluvial systems requires bridging principles to link its elements to measurable entities within the fluvial system. If force is equated with flow velocity then measurement of flow velocity becomes a surrogate for this part of the theory. But how can flow velocity be identified and measured? Flow velocity will vary across the river profile, so there needs to be criteria for selecting precisely where flow is measured. Watching a river you may notice that flow need not always be directly downstream. River flows swirl and are deflected forming eddy currents of various sizes. Again, agreement is required concerning where is an appropriate point or section of river to measure flow. The instruments used to measure flow can vary from an orange and a stopwatch along a measured reach to an acoustic or a laser Doppler device. How can we be sure that these instruments are measuring the same thing – river flow? Instruments such as a laser Doppler flow monitor measure movement based on specific physical principles about how reality works. In other words, our theory of river flow as a force is being measured by an instrument that is itself designed and developed on the basis of a theory about reality. If the principles upon which the instrument are based are wrong then the measurement of flow is flawed.

All observations are theory-ladened to some degree as Rhoads and Thorn (1996b) noted. The question is, does this theory-ladeness matter to the final identification and testing of the theory at the level of interest to the physical geographer? Identifying that laser Doppler devices are dependent upon a theory about how reality works developed within physics is something inherent to all instruments of this type. The theories associated with the production of laser light, and the interaction of the laser with the environment, are all assumed to be correct. Without this assumption there could be no assessment of theories developed at the level of the entities of interest to physical geography. We would, instead, be continually questioning and testing the physics and chemistry of our methods.

Bogen and Woodward (1988) provide an interesting spin on theory-ladeness, a viewpoint developed by Basu (2003) in relation to the work of Priestley and Lavoisier on oxygen. They suggest that, contrary to popular belief, theories do not make predictions or explanations about facts. Instead, they believe that a distinction has to be made between data and phenomena. Data have the task of providing evidence for the existence of a phenomenon and are relatively easy to observe. Data themselves, however, cannot be predicted or explained that well by theory, data just are. Phenomena, on the other hand, are detected through the use of data, but in all but the most exceptional cases, are not observable. Data stand as a potential indicator of a phenomenon, but not necessarily just one phenomenon. Data could be used to identify many different phenomena depending on what the data are thought to detect. Phenomena are viewed as being generated by establishing a particular experimental design. The design is produced to identify and monitor the phenomena. There are a limited number of circumstances that will permit the phenomena to emerge from the experimental situation. It would be expected, however, that phenomena would emerge once the limited circumstances had been obtained. Data involve such disparate and numerous combinations of factors, often unique to the experimental situation, that no general theory can be constructed to explain or predict data. Phenomena have stable and repeatable characteristics that can be detected, provided the appropriate circumstances are present, by a variety of different types of data. The different types of data need to be produced by the phenomena in sufficient and appropriate quantities to enable their detection and interpretation within the theory being assessed. Data cannot, therefore, be produced by any means, they need to be produced within a context appropriate for the assessment of a phenomenon. This means that data production is informed by the theory under assessment and will be restricted in its location and timing of production as dictated by the phenomenon under study.

To illustrate this point, we could reflect further on the lake acidification case study discussed earlier in this chapter (Battarbee *et al.*, 1985), which provides an example of a piece of research that was carefully designed in order to allow the data set to reveal particular phenomena. The study was concerned with identifying the cause(s) of recent (*c.*200 year) changes in lakewater pH. The study site selected, Loch Enoch in Galloway (southern Scotland), was located on granite geology, therefore the lake would have a very low buffering capacity and would respond more prominently to changes in pH, whether driven by natural or anthropogenic processes. The lake was located within a non-afforested catchment (afforestation was identified as one causal factor of lakewater pH change). By eliminating the 'afforested catchment variable' from the research design they simultaneously falsify this variable as a causal factor in their particular study (the lake underwent a pH change even though afforestation had not take place in the catchment) and narrow the circumstances in which the desired phenomenon (change in lake pH) could occur. The proxies selected (diatoms, pollen, trace metals) were capable of generating specific data about the phenomenon under study (changes in lake pH in the case of diatoms), the competing hypotheses under investigation (anthropogenic acid deposition (inferred from the co-variance of lake pH and trace metal concentrations) and land-use change (inferred from pollen assemblage changes)). The phenomenon emerging from the data set, that of lake pH change co-variant with a rise in sediment trace metal content since *c.*1850, is necessarily restricted to the context of this data set; indeed the authors point out that the results do not extend beyond the similar lakes of the Galloway region.

If the phenomenon can only be isolated in the laboratory, then only data produced within the experimental confines of the laboratory will be considered as relevant. Production of such data may involve long processes of establishing adequate control, techniques for ensuring data quality, as well as alteration of the data to a form amenable to appropriate statistical analysis. This theme of evidence and its construction will be taken up again in Chapter 5.

Physical geography as historical science

There has been a tendency, both historically and currently, to try to set the working practices of physical geography as a separate type of scientific explanation: as an historical science (e.g. Gilbert, 1896; Chamberlin, 1890; Johnson, 1933; Mackin, 1963; Leopold and Langbein, 1963; Frodeman, 1995 and Cleland, 2001). Even some practitioners have assumed the inferior status of an historical science. Andersson (1996), for example, in relation to explanation in historical biogeography, stated that the explanatory structure of mathematics, physics and chemistry was clearly the most prestigious model of explanation with the greatest explanatory power, but one hardly applicable to historical biogeography.

Johnson (1933) identifies seven stages of investigation in historical sciences (Table 2.2). Within each stage he believes it is vital to undertake detailed analysis. By analysis he means detailed and thoughtful testing of the basis of each investigation at each stage or 'tracing back each part to its source and testing its validity, for the purpose of clarifying and perfecting knowledge' Johnson (1933, p. 469). His preferred method for studying the environment is by the use of multiple working hypotheses, through which different ideas can be tested and analysed (see Chapter 5 for discussion of multiple working hypotheses).

The basis for distinguishing the practice and explanations developed in physical geography, whether it is in the guise of ecology or geology, from that of 'traditional' or experimental science, lies in the subject matter of study and the explanatory framework sought. An historical science is not necessarily concerned with generalities, instead it is concerned with explaining particular events, both present and past.

> Similarly, in geology we are largely interested in historical 'individuals' (this outcrop, the Western Interior Seaway, the lifespan of a species) and their specific life history. It is possible to identify general laws in geology that have explanatory power . . . but the weight of our interest lies elsewhere.
>
> (Frodeman, 1995, p. 965)

In explaining specific individuals and events, reference may be made to general laws derived from 'hard' sciences, but they will be used as part of the logical arguments

Table 2.2 *Classification of investigation based on Johnson (1933).*

Stage of Investigation	Role of Analysis
Observation	Identify all facts bearing on problem Facts assessed for their relevance to the problem and exclusion of irrelevant facts Avoidance of incomplete observation and interpretation within observations made
Classification	Grouping of relevant facts based on fundamental characteristics
Generalisation	Legitimate inferences induced from the facts themselves (p. 477) Inferences must grow out of a sufficient number of facts – triangulation?
Invention	Facts and generalisations used as basis for invention of as many explanations as possible Need to have specific deductions derived from them – testable hypotheses Mental assessment of hypotheses
Verification and Elimination	Deduces what features should characterise reality if surviving hypotheses correct Verification with existing facts confirms hypothesis Verification meaning to have shown hypothesis to be competent to explain certain facts – not as true
Confirmation and Revision	Direct observations to produce new facts to assess remaining hypotheses
Interpretation	Although a single hypothesis may survive, it is not necessarily the answer for all types of a phenomenon Present interpretation is a 'highly probably (sic) theory, rather than a demonstrated fact.' (p. 492)

concerning the specific individuals or events. The 'laws' will not be *the* explanation of the specific, instead they will form part of the explanatory framework.

The historical scientist has observed phenomena as their focus of study but without the luxury of observed causes (Cleland, 2001). In experimental sciences, causes are observed, often under controlled conditions, and phenomena can be observed as produced by them, under the controlled conditions. Historical science has to deal with what Cleland (2001) calls the asymmetry of localised events. An event such as an eruption produces a multitude of effects of which only a few need to be identified to infer the eruption has occurred. A single effect cannot, however, be used to infer the cause. There are many possible and plausible causes that can be linked to a single effect (see Chapters 4 and 6 for further discussion). Nevertheless, despite these issues, an historical science approach can generate unique insights into the workings of nature. In the wonderfully entitled paper *Coaxing history to conduct experiments*, Deevey (1969) effectively bridges the gap between the seemingly disparate approaches of experimental and historical science by viewing history as a series of experiments. He argues that history has provided some essential 'experimental' conditions, and it is the job of the historical scientist to consciously seek out these experiments. He bases this view on the premise that where time is required to see a result of an experiment, there is no substitute for history. As an example, he considers the evolution of ecosystems, and their community stability during and following disturbance in the Late Pleistocene. 'The approach is to regard historical *disturbance* as a quasi-experimental way of stressing systems to infer the nature of their stability' (Deevey, 1969, p. 40). By considering the Late Pleistocene time interval, the agents of disturbance are limited to human activity and climate change (tectonic and volcanic activity was considered unimportant owing to the short time interval under study). Careful analysis and interpretation of pollen assemblage data, as preserved in lake sediments and peat bogs for example, can be used to provide insights into terrestrial ecosystem response to disturbance; this experiment after all had been played out multiple times (orbital and suborbital climate fluctuations, localised incidents of anthropogenic deforestation) in many different locations involving many different types and combinations of ecosystems. Of course, this view could be extrapolated throughout the historical sciences; history has performed an innumerable set of experiments that are waiting to be 'consciously sought and carefully attended to by the experimenter [the historical scientist]' (Deevey, 1969, p. 40).

Historical sciences have developed strategies, amongst which is that of multiple working hypotheses, with which to whittle down the number of cause and effect relationships. Historical science searches for the most plausible set of cause and effect links as the explanation for a phenomenon. Historical science looks for what Cleland calls the 'smoking gun', the piece of evidence that unequivocally links a specific cause to a specific effect. Failing a singular, decisive piece of evidence, historical science instead focuses on identifying a subset of effects, or traces as Cleland calls them, which can be viewed as evidence that only a specific cause could have produced them. Tradition and experience will identify the small subset of traces or proxies (indirect indicators of former climates or environments) that usually provide information for this judgement as noted in later chapters.

Causation within historical science is recognised as a complicated and potentially complex affair (Inkpen and Wilson, 2009). Sets of traces, or proxies, are compiled to construct a picture of the past and the manner in which past events caused effects in the past and resulted in the present state of affairs. It is often unclear where causation lies within this picture. The bursting of a glacial dam might produce a catastrophic flood that alters the course of a river and is a major formative event in the development of a number of landforms. It may be assumed that by identifying the evidence of the dam burst, then the cause of the changing river course or slope profiles of the catchment have also been identified. The dam burst may have triggered the effects, but is it *the* cause? The flood of water would have had a lesser impact had the river channel been highly constrained by existing geology. Likewise, a large, flat, floodplain with few steep slopes would not be affected by the catastrophic flood in the same manner as a newly deglaciated landscape with unconsolidated moraines scattered across it. In other words, the context within which the trigger event occurs is as important as the trigger event itself in any explanation. Simpson (1963) noted that the immanent or ahistorical processes that experimental science is good at identifying need to be considered with the configurational aspects of a situation to understand or explain reality. Configurational aspects refer to the series of states that have uniquely occurred through the interaction of ahistorical processes and historical circumstance – the context. Without an understanding of both, causation may be mistakenly attributed to a single event. The trigger itself is not sufficient to cause the effect; the trigger is only a cause because of the specific configuration of reality it occurs within.

It is worth noting that historical science also has a major problem with identifying the nature of the entities or individuals it is trying to study (Hull, 1976, 1981). Individuals or entities do not appear fully defined and fully formed for study. A set of characteristics are required to differentiate the individual or entity from its context. Furthermore, the individual or entity may alter in these characteristics over time, as its individuality is degraded and becomes increasingly undifferentiated from its context. For example, as a mass of material moves down a slope, at what point is that mass no longer a part of the slope? When does it become an individual in its own right? Alternatively, as soil properties change moving across a landscape, at what point does one soil type become another? Although there may be standardised classification criteria for identifying soil types (Chapter 3), is a distinct value at which a soil is one thing and then another appropriate to understanding reality?

Summary

Physical geographers study a reality that they assume is external to themselves and that is capable of study. Although they may never have absolute knowledge of this external reality, they can organise and use information to argue and come to an agreement about the nature of their representations of reality. Logical positivism believes that it is possible to sense reality as it is and that it is possible to discover real laws about real entities through sensory information. Research is directed towards proving reality to be as we believe it to be. Critical rationalism takes a more critical view of our search for an understanding of reality. Critical rationalists view the testing and falsifying of ideas as the essential feature of scientific research. Knowledge is never certain or absolute, it is only the ideas that we cannot disprove that we accept as true for the time being. Within both philosophies deductive and inductive reasoning are used to reason about reality. Deduction is the only logically consistent method of argument, but unless it is connected to reality via empirical assessment, it can produce invalid arguments.

Critical realism highlights the differentiated and stratified nature of reality, the asymmetry between the real and the actual. All we can observe is the actual, but this is a pale reflection of the structures and mechanisms that underlie reality. Assessment of ideas is more problematic in critical realism, but the testing of hypotheses about reality is still a central means of understanding. Pragmaticism emphasises the importance of the mental image the researcher has of reality in understanding that reality. Any testing of ideas makes use of entities as signs, as signifiers of ideas and divisions of reality. The interpretative network of signs of which the entity under investigation is a part is not a closed network. Relations are always open to reinterpretation and entities can undergo renegotiation. The object and subject of investigation cannot be easily separated.

Understanding reality increasingly becomes a dialogue between the socially embedded researcher and reality. The ideas of the researcher are guided by theory, as are the means by which reality is investigated – the methodologies employed. Theories provide a framework for deciding what to study, how to study it and how to interpret the outcome of the dialogue.

Chapter 3

Entities and classification

Introduction

Entities, which are vital in any study, are often viewed as unproblematic and distinct divisions of reality. Traditional practices, entrenched theories and increasingly advanced monitoring technologies can all conspire to make entities appear unproblematic. Entities form the basis for the units upon which theories, and their translations to assessable hypotheses, are based. The entities studied, such as rivers, slopes, landslides and soils, indicate how researchers believe reality operates; how it is divided up, and by what processes divisions are made. Despite their pivotal role, entities remain relatively poorly conceptualised. To many, problematising the status of entities appears to bring into question the whole process of scientific study. Questioning entities, however, is a key part of the dialogue between researcher and reality. What is being studied is being continually redefined, renegotiated and refined. The whole process of research results in an often slow, occasionally rapid, invisible renegotiation of an entity. The same name may be used, but the connotations associated with an entity change.

It is clear that physical reality does not change as paradigms change. It is even likely that the property-based definitions of mountains do not alter. So what does change? What changes is how the entity or individual is defined and the classes into which these individuals can be divided. Therefore, a central concern has to be what do we mean by change in entities? Are we discussing a physical reality or a mental construct? Unfortunately, the answer is both. This is because physical reality can only be grasped via our theories. If these alter, then what we believe physical reality to be will also alter, even if it appears that the entities remain the same physical things. Entities become defined and entrenched within particular theories, which in turn provide the basis for identifying new properties that further help to define the entity, as well as guiding how the entity should be measured. In other words, new theories renegotiate entities in their own image.

Physical geographers have often engaged in redefining and reconceptualising the nature of the entities they study. They would probably generally view this engagement as resulting in an improvement of their understanding of reality rather than as a process of renegotiation. This chapter tries to address the issues above by looking at, first, the nature of entities, what they are and how they are defined, as well as why they are important. Of importance in this discussion is the relationship between entities and kinds. Kinds are linked to the issues of classification and standardisation of reality. The act of classifying and standardising gives an impression of a fixed, unchanging and therefore objective reality.

What are entities?

Philosophically, this is a tricky question. Entities are the units we believe exist in reality. They are the things we try to study; how they behave and why they behave in that manner is what we try to explain. Whether they really exist as we think they do, however, is a difficult issue to clarify. We may, for example, study a landslide and try to explain its behaviour by reference to smaller parts. By focusing on the component parts, does this mean that the landslide is not really the focus of study or the basis of explanation? Are not the component parts the focus of study (the things that really exist) and hence represent the 'real' entities of study? Could we go even further and argue that the real basic entities can only be described through their fundamental physical and chemical properties, with physical geographers merely studying aggregates of these basic entities? This next section will highlight why, within physical geography, such a reductionist view of entities, and the reality they represent, is misplaced.

A key characteristic of an individual entity is that it is assumed to exemplify a distinctiveness through having a particular property or attribute or set of these attributes (Loux, 1998). However, viewing entities as isolated islands of distinctiveness is not of great use in explanation. Entities need to be viewed as signs or as representative of something else, as illustrations of universals, for example, to be of use in explanation.

There are at least two possible views of entities (Loux, 1998). One view is that reality is made up of universals. These universals are real entities that find expression, at least partially, in entities we find in our 'real' or experiential world. Such universals can be exemplified simultaneously by different objects, as there need not be a one-to-one correspondence between universals and experiential entities. Universals define the properties a thing has and its relationship to the kind to which it belongs. Mudslides have a set of properties by virtue of being mudslides. These properties are universal, possessed by any entity defined as a mudslide. The mudslides we observe are not 'pure' however. The realisation of the full set of properties may be polluted by the nature of the material the mudslide occurs in, the specific environmental conditions it occurs under and so on. We know that these properties are not fully expressed because we have in mind the ideal type, the universal mudslide, which exists in reality but that only finds expression in the pale realisations we can observe.

An alternative, second view of entities is that reality is only made up of individual entities themselves. These entities are not examples of some underlying universals. Instead, each entity has properties that may or may not be the same as properties possessed by other entities. For example, in this view, two mudslide events may occur that possess the same velocity of movement, that are composed of roughly the same material and seem to behave in a similar manner as far as we can identify and measure their behaviour. These mudslides have similar properties because we say they do. Possession of similar properties is not an indication of any relationship between the two entities, of any prediction of behaviour of one by the other – this possibility would imply that both entities are linked in some manner to a universal. Both views of reality do, however, accept that entities can be structured and complex in nature, and not the simple basic units of reality a more reductionist view might envisage.

Properties that are the hallmark of a distinct entity can, however, be considered in different ways. Any entity could be viewed as an entity shell devoid of any properties, the so-called substrata view. Properties are independent of the entity and not necessary for the entity to exist. An entity exists and properties fill it, but the properties do not define the nature of the entity. The properties, despite being associated with an entity, do not define it. Almost diametrically opposed to this view is the concept of an entity as a bundle of properties. There is no empty shell to fill. Instead, it is the association of properties or attributes that define the entity. An entity has no physical existence outside of the potentially contingent association of properties. Without properties the entity melts away, it does not exist, it has no substance. Both these views have severe metaphysical problems (see Loux, 1998) and within physical geography neither would be able to enhance explanation. The first assumes the existence of an entity without properties, the very things we identify, define and by which we measure an entity. The second provides no solidity to a form, just viewing an entity as, at the extreme, a contingent and potentially random association of properties. An understanding of reality based on either would be very limited.

Entities and kinds

A way out of the explanatory dead end of the above views of entities is provided by the concept of substance or essence. The idea has been traced back, as much of philosophy can, to Aristotle. The central tenet of the idea is that entities can be seen as irreducible. Irreducible entities are not merely the sums of their parts, they are more than this. When put together the whole has properties grounded in universals that refer to the essential nature of the substance of the entity. It is these properties that define an entity. This means that not all properties associated with an entity are of equal importance for the essence, the substance of the entity. Without a limited subset of specific properties the entity could not exist or be identified. These properties define, or rather permit, the researcher to identify the substance of the entity. The properties are not, however, filling an empty shell. The properties exist, take the values they take, and enter into the relations they do because they are part of a specific entity. The relationship between entity and properties is mutual and indivisible.

A key universal is that of a kind. Kinds provide an identikit of the essence of an entity. Membership of a kind defines those properties that should be present in any entity of that kind. Non-essential properties do not alter an entity's membership of a kind, but they can mean that entities of the same kind need not all have properties in common, nor essential properties with the same values. Kinds can be thought of as templates, outlining the basic contours of what it means to be an example, a member of a kind. Kinds can also form nested hierarchies of more general or more specific essences depending on which way you move through the hierarchy. Animals, for example, could be seen as a general kind of which apes are a more specific kind and humans an even more highly specific kind. Each more specific kind retains the general properties of the more general kind but is differentiated by other properties that do not contradict or prohibit their membership of the more general kind. Linnean classification in biology could be seen as a classic example of nested kinds.

Providing a template for the essence of an entity is fine, but does it aid explanation? If you had an entity of a certain kind, you would expect it to behave in a certain way. Membership of a kind defines the relationships possible, the essential properties, and, potentially, how an entity can change. But is the kind then the appropriate focus of explanation? Could not a geomorphological feature, such as a slope, be composed of other smaller kinds such as soil? Are these smaller kinds more fundamental than the slope? Are they more appropriate as foci of explanation? Could you keep this reductionist regression going until you reach 'real' kinds, often referred to as natural kinds?

Rhoads and Thorn (1996b) highlight the significance of the object rather than the method of study for geomorphology. Rhoads and Thorn (1996b) discuss whether landforms are natural or nominal kinds. Natural kinds they view as having some objective and real nature in reality, whilst nominal kinds are humanly constructed artefacts (Schwartz, 1980). If natural kinds exist, their objective nature could imply a superior status in any explanation. Explanation involving natural kinds would be viewed as involving real things, reality as it is, rather than a potentially subjective and changeable human artefact. Studying an entity as a natural kind has distinct advantages. Researchers can assume that they are studying the world as it really is and so obtain objective, independent information. Investigation of natural kinds derives the essences of 'real' objective entities, their generative mechanism and causal powers (Harre and Madden, 1973; Wilkerson, 1988; Putnam, 1994). Powers of an entity exist independent of its context and who is studying it.

The identification of a natural kind from a nominal kind becomes important for explanation if only natural kinds are seen as really existing. Establishing some criteria or practice for distinguishing natural and human kinds becomes a vital task. Putnam (1973) and Kripe (1980) claim that empirical research provides the basis for naming and assigning natural kinds. Li (1993) suggests that this merely moves the question of kinds from the realms of philosophical debate to scientific debate without really tackling the basis of the question. Schwartz (1980) similarly holds that natural kinds are defined by traits that are discoverable empirically and tested for by his counterexample test. Scientists identify which traits are associated with natural kinds through study of individual entities. These entities are, however, studied in a manner consistent with their being the member of that kind. This somewhat circular argument makes it difficult to break out of an empirical bind.

Unfortunately, for physical geography, the status of entities such as rocks, rivers and even the events that trigger landslides, as natural kinds is debatable. Aristotle, for example, only gave the status of natural kinds to a limited range of entities (Loux, 1998), the elements of his day. Contemporary fundamental quantum physics might claim the same status for its entities, but the changing nature of what has been seen as fundamental or elementary entities over the last 2,000 years seem to count against a view of a once and for all answer to the delineation of a fundamental 'natural' kind. Although the existence of geographic entities is not denied, except in extreme interpretations (e.g. van Inwagen, 1990), their explanatory status is often viewed as inferior.

Dupre (1993) and Shain (1993) suggest that the definition of natural kinds is dependent upon the context of enquiry. Li (1993) argues that in naming a natural kind there is always a vagueness at what, precisely, the process of naming is directed. It is only possible to know natural kinds through instances, individual entities that exemplify the kinds. There is no direct access to a natural kind. Naming is based upon experience, both individual and collective, and not upon some absolute knowledge of reality. Identification and naming is based upon comparison of entities. As the process of comparison can be virtually endless, it is for all practical purposes never complete or capable of completion. This means that the absolute assignment of an entity to a kind can never be certain.

> I have argued that, because we have no direct access to a natural kind as we do to an individual object, when we name a natural kind exactly, what the kind is cannot be determined without further focus. The process of further focusing can never come to an end.
>
> (Li, 1993, p. 276)

Nominalists, according to Hacking (1983), do not deny the reality of an external world containing entities which interact. They believe, however, that we impose our classification, our own divisions, upon this reality. Hacking (1991) suggests that kinds become defined and important when their properties are important to the individuals who want to know what entities, and by extension, kinds, do and what can be done to them. Practice and experience become central and important factors in the process of identifying kinds.

> When we recognize things to use, modify or guard against, we say they are of certain kinds. Singular properties are not enough. Realizing that a thing has some properties or stands in certain relations prompts belief that it is of a certain kind, i.e. has other properties or stand in other relations. Kinds are important to agents and artisans who want to use things and do things. Were not our world amenable to classification into kinds we cognize, we should not have been able to develop any crafts. The animals, perhaps, inhabit a world of properties. We dwell in a universe of kinds. . . . Natural kinds, in short, seem important for *homo faber*.
>
> (Hacking, 1991, p. 114)

Hacking (1999) identifies two types of kinds that may prove a useful typology in geography. Indifferent kinds have no reaction to being named. A limestone rock does not respond to being classified as a limestone rock. Its nature or essence does not alter nor does it begin suddenly to behave as a limestone rock when it did not before. This is not to say that we, as investigators, do not interact with the rock in a particular manner because we have defined it as a limestone rock. A stonemason might work the limestone in a particular way and expect a particular response. Scientists may test for the presence of limestone by watching for bubbling from the rock surface after the application of hydrochloric acid. Neither of these behaviours provokes a response from the rock independent of our interpretation and action upon it.

Interactive kinds, on the other hand, react to the act of classification. An individual defined as a refugee becomes embroiled in a set of relationships that operate because of that act of classification. The individual reacts to that classification and responds in a manner permitted by or in keeping with that classification. The act of classification provokes a response. Most, if not all, kinds in physical geography could be thought of as indifferent kinds. There may be an implicit assumption that indifferent kinds are somewhat more 'real', more stable than interactive kinds and therefore a 'better' basis for explanation. Even indifferent kinds, however, are both social and 'real' (Hacking, 1999).

The basis for identifying and researching kinds is the dialogue between reality and the researcher. Extraction of properties is always made for some purpose. The distinction is not between some real, indifferent natural kind and some interactive socially constructed kind. There is no distinction to worry about. Trying to establish a distinction is irrelevant. Goodman (1978) identifies relevant kinds as those defined for some purpose as relevant by the user. Goodman emphasises that relevant kinds assume an external reality with real entities. This reality is, however, only accessible through our actions, guided by our intents. Why we divide reality up, and the entities we make, are for our own relevant ends. There is no indication and no way of knowing, whether our divisions of relevance match 'real' divisions.

Kinds in physical geography are always relevant kinds derived from a time- and place-specific dialogue. Their contextual nature makes them no less real in the framework of study of which they are a part.

Case Study

Species as natural kinds

The status of species as a natural entity (Ruse, 1987) for biological classification has been questioned by a number of researchers. Dupre (2001) suggests that the species do not effectively function as a unit for both evolution and classification in general. He makes the point that a unit of evolution needs to be an individual, whilst a unit of classification by definition has to be of a kind. It is the individual organism that is in competition with others of its kind as well as eking out a life from the environment. According to Dawkins (1978), the unit of evolution is the gene that provides the basic building block for evolution to occur. Organisms are merely differentiated shells surrounding genes. The interdependence of organism and gene is just one problem with this restrictive view of the basic unit of evolution. According to Dupre (2001), evolution occurs when a set of properties that characterise individuals in a phylogenetic lineage change over time. The lineage is the unit of evolution, not some entity at a single point in time. If this is the case then what properties alter and how does their alteration not change the essence of that species or gene? Both for species and genes the temporal dimension raises problems for identification and classification. The idea that a species can be the same thing throughout its evolution implies that the same species, or rather aspects of it, exists at all stages of its evolution. Such a view implies that each current species is somehow inherent in its past evolutionary lineage.

Levins and Lewontin (1985) and Lewontin (1995) highlight that the organism may be the unit of evolution, but it is not necessarily a passive unit. A classic view of evolution implies that the environment sets problems and organisms find solutions. The environment is fixed; organisms are malleable. The organism is a passive respondent to changes in the environment. The advent of genetics may seem to imply a more active role for the organism, but instead it has emphasised the internal, inherent characteristics of the genes. The organism is still a passive unit – the genes alter so the organism has to. The organism becomes almost an infinitely deformable conduit for the external influences of the physical environment and the internal influences of its genetic make-up. Levins and Lewontin (1985) and Lewontin (1995) instead suggest that the organism is an active motor of change and, at least partially, a determinant of its own destiny. They base their claims around a number of key points. First, organisms determine what the relevant characteristics of the environment are for them. The concept of a simple niche as a model of reality is difficult to sustain. Adaptation to fit a niche requires defining the niche before the organism exists. In a similar vein, organisms can create their own microenvironments. As Levins and Lewontin (1985) note, even without manipulation of their surroundings, all terrestrial organisms have a boundary layer of warm air resulting from the metabolism of the organism. Second, organisms alter the external world that they interact with. Plants alter their substrata by their activities (e.g. local alteration of soil structure and chemistry). Third, organisms interpret the physical signals they receive from the physical environment. Any environmental change is mediated to the organism through its senses and its response will depend on what it senses and how it interprets this change. Fourth, organisms can affect the variability of the physical environment spatially and temporally. Changes in air temperature may be dramatic, but an organism with feathers does not sense anything other than extreme changes in temperature. Finally, it is not the environment that defines the traits to be selected by evolutionary processes, it is the relationship between the organism and the environment that defines these. The organism is an active participant in the environment and it is how the organism relates to the physical environment, to organisms of other species and to other organisms of the same species that determines which traits are of use and which are not. There is no predefined set of traits that are inherently better than some other set of traits; it all depends on context. The above points illustrate that the unit of evolution is difficult to isolate from its relations. Likewise, the organism as a member of a kind needs to be thought of as an active rather than passive agent. Its properties are context-dependent and subject to change, although their potential existence may be present in the kind of which they are a member. Activation of these properties is not inevitable, it all depends on the relations an organism has in a specific context.

The above discussion implies that classification of species is fraught with problems. Dupre (2001) suggests that classification, when associated with a central theory, can produce a 'real' classification of reality, even if it is, he believes, a weaker claim than traditionally assumed for natural kinds. Kinds themselves, however, are not the sort of things that exist in reality. Instead it is the members of the kind that undergo the effort of existing in reality.

Case Study

Magnitude and frequency – entities out of context

Wolman and Miller (1960) outlined the significance of magnitude and frequency of events in geomorphology by abstracting events from their context of operation. Measuring an input, an event, into the system and then its immediate output or effect, they could construct a relationship between the two that was meant to hold for any fluvial system anywhere. Events were defined as discharges of given magnitudes, whilst output was defined as the sediment load carried by the river. The theory they developed, or 'truism' as Haines-Young and Petch (1986) describe it, was that middle-size events (normal events) did the most work in a catchment (Figure 3.1). Small events were frequent but of insufficient power to move material. Large events moved a great deal of material but were not frequent enough to move material with sufficient regularity to sculpt the landscape. It was the middle-size events (bankfull discharge events in their study) that occurred with sufficient frequency and force to move material around and out of the catchment so as to sculpt the landscape.

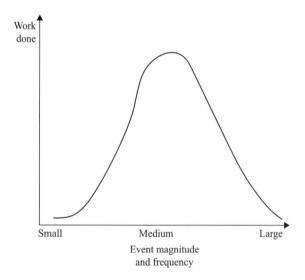

Figure 3.1 Magnitude/frequency relationship.

Lane and Richards (1997) noted that this initial description of the relationship between magnitude and frequency relied heavily upon the idea of a 'normal' operating mode for the fluvial system within a catchment. The idea that the fluvial system tended towards an equilibrium form with a 'normal' mode of operation was an important basis for legitimating the abstraction of events and effects from their context. Abstraction was permitted, because the events and effects were illustrations of this normal mode of operation. Event characteristics were reflections of an ideal event type that had a specific and contextless relationship to an equally ideal effect. This meant that the relationship between event and effect could be modelled as a simple input-output relationship that could be transferred between fluvial systems. The magnitude-frequency curve derived could be applied to any system regardless of its environmental context.

Subsequent development of the magnitude-frequency concept by Wolman and Gerson (1978) gave some recognition to the context-dependence of the abstracted relationship. They introduced the idea of effectiveness to describe how different landscapes may respond differently to the same event, the same input. Effectiveness related the magnitude of the change caused by an event to the ability of the landscape to restore itself. Both parts of this equation were context-dependent. Within a temperate environment an event of a given magnitude will have an impact of a specific magnitude. The temperate environment will tend to recover from this impact and move back towards its 'normal' state. In a semi-arid environment, the impact from an event of the same magnitude will be greater and the time required for recovery will be longer, possibly even longer than the time intervals between events producing the same impact. Although context is now introduced into the analysis, it is still a very limited view of context. Likewise, the system still retains a movement towards equilibrium or 'normality' as its functional goal.

Lane and Richards (1997) highlight the need to distinguish between the 'immanent' ahistorical processes and the 'configurational' (Simpson, 1963). The former are general processes that always occur under specific conditions, whilst the configurational represent the result of the historical interaction of the immanent with particular historical circumstances. This mirrors the complicated relationship between event magnitude-frequency and impacts. The relationship can be abstracted and general properties derived. The relationship is a generalisation, often a statistical caricature of complicated and temporally distinct relationships. The relationship cannot be understood as an ahistorical process. Generalities are only derived from particular circumstances. Without knowledge of these circumstances the generalities cannot be related back to reality. General properties are only discernible because the processes that cause them operate within specific contexts derived by unique historical sequences of change.

Semiotics as put forward by Baker (1999) has a similarly blurry image of entities and recognises their indeterminate status. Entities have to be viewed within the web of signs of which they are a part as noted in Chapter 2. Entities, and by extension kinds, become contextually defined rather than objects defined independently of study context.

> As a thing it merely exists, a node of sustenance for a network of physical relations and actions. As an object it also exists for someone as an element of experience, differentiating a perceptual field in definite ways related to its being as a thing among other elements of the environment. But as a sign it stands not only for itself within experience and the environment but also for something else as well, something beside itself. It not only exists (thing), it not only stands to someone (object), it also stands to someone for something else (sign). And this

'something else' may or may not be real in the physical sense; . . . Divisions of things as things and divisions of objects as objects are not the same and vary independently, the former being determined directly by physical action alone, the latter being mediated indirectly by semeiosy, the action of signs . . . Divisions of objects as objects and divisions of things as things may happen to coincide . . . But even, when they coincide, the two orders remain irreducible in what is proper to them.

(Deely, 1990, pp. 24–25)

This quotation provides a useful perspective on the semiotic or pragmatic realist interpretation of entities. An entity takes on a meaning and is only constituted because of its relations in a triad of sign, interpretant and interpreter. The entity exists as a thing, an object and a sign simultaneously and each of these elements is interdependent and uninterpretable without the others. The sign, the focus of study in semiotics, is the 'something else' of the quote. The sign is constructed through interpretation by an interpreter of an object. If the object and thing are identical the sign is an interpretation of reality as it is. The coincidence can never be known for certain, but it would also be inappropriate to assume its presence meant an object could be reduced to a thing or even vice versa. The nature of a thing is associated with its relations to a physical reality, relations beyond the knowable for a researcher. The thing is wholly or partially related to an object that is constituted by relations to the researcher. This latter set of relations means that the object is constituted via signs, it stands as an outcome of interpretation. This means that its very nature is different from that of a thing, the reality, that it is a representation of. Any knowledge a researcher can glean about reality is via the object and so only knowable via signs, via interpretation. Knowledge without relations to signs is not possible.

Entities are not reducible to some common baseline of reality, they are only reducible to objects of study, with all the difficulties of being part of a dialogue that this implies. In some ways, this interpretation of the individual entities is similar to the relational view of entities put forward by Whitehead (1929). He understood entities to be capable of distinct definition but only by virtue of an entity's relations to other entities and, in turn, their relationships to the wider whole of reality. Any single entity at a specific time contains within it its relations to itself in the past and future. Past relations express the constraints on an entity, whilst future relations, as yet unrealised, express the potentialities of an entity. In Whitehead's view, all the relations that come to define an entity in the present are the constraints of its past, and explain why it is as it is. The uncharted future of an entity is not random or completely expansive. An entity's future is constrained but not singularly determined by its past (what it is). The nature of an entity defines its possible futures, its potential relations. In this way, an entity embodies both what it is, what it has become and what it could be or is becoming. There is, however, no need for an entity's nature to be rigidly fixed or stable over time. Entities can vary in their spatial and temporal stability. Whitehead's vision of entities as observer-defined abstractions, solidifications, of the flow of processes, captures the distinct but ever-changing nature of an entity. Entities require flows of processes, their relations with other entities and the wider whole of reality to continually reproduce themselves. Stability of these relations provides the appearance of solidity, of illusionary permanence. Semiotics, with the continual creation and destruction of networks of signs that constitute an entity, is a view in a similar vein.

The semiotic approach means that both entities and kinds do not need to have any physical reality, although they will have relations to this reality. In semiotics and pragmatic realism there are no natural kinds or entities, neither exists independently of an interpreter.

> And the metaphysical categories that emerge through this endeavour yield an understanding of nature or the dynamical object which, as an indefinitely rich evolving continuum that must be made determinate for our awareness by the manner in which we 'cut' into it, cannot provide the basis for final ultimate knowledge. Nature or the dynamical object, with its qualitative richness, lawful modes of behaviour, and emerging activities, constrains our interpretations, pulling them or coaxing them in some directions rather than others. It answers our questions and determines the workability of our meaning structures, but what answers it gives are partially dependent on what questions we ask, and what meaning structures work are partially dependent upon the structures we bring. Thus, within this interactive context of interpretation and constraint, different structurings yield different isolatable dynamical objects, different things, different facts. Truth is always related to a context of interpretation. This is not because truth is relative but because without a context of interpretation the concept of truth is senseless, indeed literally so. Knowledge involves convergence, but convergence within a common world that we have partially made, and continually remake in various of its aspects and in various ways.
>
> (Rosenthal, 1994, pp. 127–128)

As the quotation points out, the divisions of reality, the kinds studied, are of our making. We do not, however, start with a blank page. All our interpretations are constrained by an underlying and independent reality. We interrogate that reality and kinds emerge as an outcome of that interrogation, but are not existent independent of that interrogation. Entities of study, likewise, are dependent upon this interrogative process. It is of vital importance, however, to realise that the dependence of kinds, entities and objects upon interpretation does not mean that they reflect a purely relativistic version of reality. Kinds and entities are not relativistic in the sense that there is an infinity of other possible interpretations, all equally valid, as there is no basis for assessing the validity of each. The basis for assessing the validity of entities and kinds constituted is the interpretative context of their constitution. It is only relative to this context that appropriate methods of validation can be constructed. It is the interpretative context within which judgements concerning appropriateness can be passed. The interpretative context, like the entities and kinds within it, is open to continual renegotiation, to continual alteration and evolution. Many scientists would claim that this renegotiation is the convergence mentioned in the quotation towards a 'truer', in the absolute sense, version or interpretation of reality. This assumes a progressive direction to the alteration of the interpretative context of which we can never be certain. The continuation of certain entities and kinds with changing interpretative contexts testifies to their adaptability rather than proof of their ultimate reality.

A more appropriate metaphor for alteration of interpretative contexts may be that of evolution. Changes in interpretation in the interrogation are appropriate for the needs of the researcher in that time and place. These needs, and the alterations they cause, constrain the nature of further alterations in interpretation, but how is unclear, because, in the present, we work only with the singular, historically unique and constrained outcome of the past.

Classification

The above discussion of entities and kinds provides the basis for dismissing the idea that there are natural classifications in reality. Division of reality is undertaken by researchers working within a unique interpretative context with associated versions of kinds and entities. The above discussion suggests that these kinds and entities do not correspond to reality as it is but rather to reality as a useful framework for the researcher. Classification practices reflect this view of reality. Classification is based on usefulness to a researcher rather than determining the absolute structure of reality. Classification of reality, therefore, becomes a means to serve the researcher or group of researchers' ends. Classification is a research tool, like any other; it is an aid to interpretation, rather than an absolute statement about the nature of reality.

Decisions about how to divide up reality, which at heart is what classification is, require use of some property or properties associated with entities and kinds. There are two ways in which properties can be used to classify reality: by lumping things together or by splitting reality apart. Lumping and splitting are two diametrically opposed ways of looking at reality. 'Lumping' views properties as being cumulative means of combining entities into a single reality. Properties should be chosen that help in linking diverse entities. Commonalities are searched for. Splitting, on the other hand, looks for properties that enable an initially undifferentiated reality to be divided into distinct, unique parts. The two approaches to classification could be seen as bottom-up (lumping) and top-down (splitting), the former working from the level of individuals up, the latter down from the level of the morass of an undifferentiated reality. Choosing whether to lump or split can also influence the seemingly objective choices of quantification of classification. Foley (1999), for example, in relation to analysing hominid evolution, noted that multivariate statistics tend to highlight overlaps between hominid morphologies whereas cladistics operates most effectively when differences between morphologies are the focus of study.

A key question is which properties to select to differentiate reality, whether lumping or splitting. Property selection for classification may be related back to the theoretical framework used. Theory defines which properties reflect the essence of entities and which are contingent. Theory, in other words, guides the researcher and defines which properties to use for classification of the 'real' entity or kind. Ideally, properties that are essential for the entity to exist would form the basis of classification. These properties are in combination associated only with that entity and so provide a clear and unambiguous means of separating the entity from the rest of reality. Such property identification and association is rare. Often classification requires not properties unique to an entity, but properties that vary between entities. These can be measured over a range of values and entities assigned to a class based on their position in this limited continuum. Assignment requires division of the range of properties into specific zones. If a property value falls within that zone the entity is assigned to that class, if the value is outside the zone it is assigned to another. Many properties can be used in combination to provide a more complex, composite measurement scale or rather multidimensional measurement space in which to classify entities. The principles are, however, the same as for a single index. The measurement space is divided into discrete volumes. A composite value within the range of a specific volume means assignment to that class.

The EU Water Framework Directive provides a useful example of how a classification scheme is constructed and how the dialogue between researcher and researched enables refinement and extension to the classification scheme.

Case Study

The EU Water Framework Directive and the classification of surface water status

The EU Water Framework Directive (WFD) is a piece of legislation that came into force in December 2000 (European Union, 2000). The WFD aims to prevent the further deterioration of European ground waters and surface waters (rivers, lakes, transitional waters, coastal waters) and to achieve 'good surface water status' in all relevant waters by 2015. Principally, the WFD uses a collection of properties – 'biological quality elements' (phytoplankton, macrophytes and phytobenthos, benthic invertebrate fauna, fish fauna) – to assess and classify the status of surface waters into one of five categories: high, good, moderate, poor, bad. This is in addition to the other properties of chemical and hydromorphological elements. Europe has been divided into 25 ecoregions for rivers and lakes (e.g. Great Britain, Dinaric Western Balkan, Tundra), and six ecoregions for transitional waters and coastal waters (e.g. Atlantic Ocean, North Sea, Mediterranean Sea). Within each ecoregion, river basin districts have been identified by their member states, and the water bodies within each district allocated to a surface water category (e.g. rivers, lakes, coastal waters). For each surface water category, a number of different ecotypes are recognised to account for the natural variability in ecological communities. For example, differences in altitude, catchment area and geology are used to classify rivers into distinctive types, whilst lakes are classified based on altitude, depth, surface area and catchment geology. The surface waters of each ecotype are classified as being of high ecological status if the taxonomic composition (and abundance in some cases) of the various biological elements corresponds totally, or nearly totally, to undisturbed conditions (i.e. in the absence of any human impact). The value of the various biological quality elements of a specific surface water ecotype of high status constitutes the biological reference condition for that specific surface water typology. In order to implement the WFD – to classify the ecological status of surface waters, and to achieve the aim of restoring surface waters to good ecological status – we must have knowledge of the biological reference condition of each surface water ecotype (i.e. the taxonomy and abundance of specific aquatic flora and fauna) in undisturbed surface waters.

The identification of a biological reference condition for each surface water ecotype is problematic. Most of the ecoregions are located in areas with a long history of human activity and so this rules out the identification and use of undisturbed surface waters as reference sites. For example, analysis of pollen assemblages preserved in lake sediment deposits provides evidence of human activity in lake catchments from *c.*6,400 years BP in NW Greece (Lawson *et al.*, 2004) and *c.* AD 870 in northern Iceland (Lawson *et al.*, 2007). Even in remote ecoregions, such as the Tundra ecoregion (which includes the islands of Svalbard and Franz Josef Land), it is unlikely that some of the biological elements of the surface waters are unaffected by human influence. In a comprehensive study into the biological communities of arctic lakes, Smol *et al.* (2005) identified widespread species changes and ecological reorganisations in algae and invertebrate communities since *c.* AD 1850 thought to be driven by climate warming and associated lengthening of the summer growing season and other changes in limnological parameters. The authors warn that '*the widespread distribution and similar character of these changes indicate that the opportunity to study arctic ecosystems unaffected by human influences may have disappeared*' (Smol *et al.*, 2005, p. 4397).

One approach suggested in the WFD is the use of palaeoecology in identifying biological reference conditions. Diatoms in particular (representing components of the phytoplankton and phytobenthos) can be used to infer the reference condition for lakes of different ecotypes. Diatoms (Class Bacillariophyceae) are a group of microscopic algae that are abundant in almost all aquatic habitats. Diatom valves are typically well-preserved in lake sediments (and in the sediments of other depositional environments, e.g. estuaries), and the size, shape and sculpturing of their siliceous (opaline) cell walls are taxonomically diagnostic (Smol and Stoermer, 2010). Analysis of the fossil diatom assemblages, and also other fossil indicators (e.g. aquatic plant macrofossils, chironomids, fish scales) preserved in lake sediment deposits, can potentially allow an insight into the biological composition of lake ecosystems prior to significant human impact (Bennion and Battarbee, 2007). However, there is a trade-off involved when using palaeoecologial techniques to identify lake reference conditions. In the UK, for example, the ecological status of lakes (inferred using subfossil diatom assemblages) prior to major industrial and agricultural intensification from *c.* AD 1850 has been considered appropriate to define reference conditions (e.g. Bennion *et al.*, 2004), and this date has also been adopted elsewhere (e.g. in Ireland: Leira *et al.*, 2006). Even so, it is accepted that diatom-inferred reference conditions immediately prior to *c.* AD 1850 are unlikely to equate to natural or pristine states due to the long timescales in which aquatic systems have been exposed to human impact (Bennion *et al.*, 2004; Leira *et al.*, 2006). This problem is nicely illustrated in a study by Bradshaw *et al.* (2006) in which the validity of using *c.* AD 1850 as a date to define a reference state for Danish lakes is tested. A pollen and diatom record stretching back almost 7,000 years from Dallund Sø, a lake in Denmark, revealed several prominent intervals of human impact on the lake ecosystem. Moderate nutrient enrichment of the lake due to agricultural impacts during the Bronze Age (1700–500 BC) and the Iron Age (500 BC–AD 1050) is inferred. More significant nutrient enrichment is apparent between AD 1050–1536, partly due to the retting of hemp and flax (retting refers to the process of soaking plants in lake water for subsequent use in the production of fibres for cloth and rope making). Overall, there has been a cultural impact on the Danish landscape since the introduction of agriculture 6,000 years ago. Nevertheless, when employing palaeoecological techniques to infer lake reference conditions, there is little alternative to using subfossil (c.1850) phytoplankton and phytobenthos assemblages. One may argue that the early to mid-Holocene, an interval of minimal human impact, should be used to define baseline reference conditions for different lake typologies. However, this is problematic because the lake phytoplankton and phytobenthos from this interval would reflect the natural (unimpacted) status of lakes under climate conditions that were different from the present. Additionally, important changes in lake chemistry, such as a progressive increase in acidity and a change in nutrient concentrations, can occur as part of the natural ageing process of lakes (Engstrom et al., 2000), thus the phytoplankton and phytobenthos from early to mid-Holocene lake deposits would reflect lake conditions at an earlier stage in lake development.

Diatoms are also used widely to help classify and monitor the quality status of rivers and lakes for the purpose of the WFD. As with the identification of suitable biological reference conditions, there is also a degree of uncertainty involved in assessing and classifying the quality status of rivers and lakes (Kelly et al., 2009). Stream/river bed and lake littoral zone diatom assemblages are influenced by a large number of environmental variables (including light, temperature, substrate (e.g. clay, sand, gravel, vegetation), nutrient availability) as well as biological variables (e.g. community succession, competition). Consequently, diatom communities exhibit significant spatial and temporal heterogeneity (Smol and Stoermer, 2010). This raises the possibility that the true status of a water body will not be correctly identified if the number of samples used for classification is insufficient, and thus unrepresentative of the ecosystem under study. This is a very important issue because the failure of water bodies to attain a 'good' ecological status will trigger expensive intervention schemes to improve their status. Kelly et al. (2009) recommend that a number of replicate samples over two or more years are used in order to capture the range of natural variation in species populations and to reduce the uncertainty in water body quality status classification. This issue of spatial and temporal heterogeneity is also common to other phytoplankton groups. Lepistö et al. (2006) attempted to classify the typology of 32 non-impacted Finnish lakes based on phytoplankton species composition and total phytoplankton biomass. A preliminary application of the WFD lake typology criteria (altitude, latitude, catchment geology, lake basin area) for Finland's lakes identified ten types (Pilke et al., 2002); lakes from eight of these ten types were used in the phytoplankton classification feasibility study of Lepistö et al. (2006). Five lake types, rather than eight, were distinguished based on phytoplankton composition alone. The phytoplankton composition was characterised based on one sample from each lake taken during one week in mid-July 2002 and so would not have captured the spatial and temporal variability in the total phytoplankton communities. Furthermore, the geography and climate of Finland increases the likelihood of some diachroneity in the timing of seasonal phytoplankton succession, which is influenced significantly by the prevailing weather conditions. This will further confound attempts to classify the biological quality status of lakes using few sample replicates if the lakes within a given typology are spatially disparate.

The issue is more complex, however, than just identifying properties of essence. Often properties useful for the classification cannot be measured. They may be properties that exist in theory but which have no clear or tenable translation into measurable properties. In these cases, surrogate properties that are measurable and assumed to be related in some manner to the unmeasurable theoretical properties are used instead. The problem is that there is not usually a one-to-one relationship between measurable and theoretical properties. A measurable property may only reflect a particular aspect of a theoretical property. Similarly, a property may be illustrative of a number of theoretically derived properties. A simple measurable property may reflect a whole host of theoretical properties to varying degrees.

A lack of a one-to-one correspondence does not provide an ideal basis for classifying entities in an absolute manner. The implication is that all classification is conditional although some classification schemes seem to be more stable than others. The Linnean division of the living realm, as mentioned earlier, seems to be a successful and stable classification scheme. Stability does not, however, equate with reality. Just because a classification has been stable or appeared stable does not mean that we can say for certain that its divisions divide reality as it is. As with entities and kinds, classes are convenient and relevant divisions for researchers.

Classification using measured properties requires another factor, standardisation. Classification, particularly in the natural sciences, requires the repeated measurement and consequent consistent assignment of entities to a specific class. Classification would expect or demand that an entity be assigned to the same class by independent observers. Property measurement and class allocation need to be consistent across space and time to permit such comparison and assignment (e.g. as discussed in the case study above concerning the importance of standardising sampling procedures to achieve a robust classification of biological quality status for implementation of the EU WFD). Without confidence in the classification procedure providing consistent assignment, researchers could not be sure that they are studying and discussing the same entities and kinds. Without this confidence, interpretation of phenomena would not be a focus of analysis, rather it would be the validity of the entities used that would consume the attention of researchers.

In this context, the process of standardisation of property identification, measurement and use for assignment ensures that definitional quality is maintained or guaranteed. Standardisation, however, requires the implementation and co-ordination (convergence and alignment in ANT terminology) of a complex social and material network. Indeed, vast volumes of papers have been written to justify a particular classification system and the particular methods of measurement required to implement it. The example of soil classification provided at the end of the chapter illustrates how classification schemes require detailed and often committee-based agreements concerning the precise methods to be used and the interpretation of results. Such concerns are not minor details amongst important theoretical or philosophical discussions. Given the central importance of standardisation to the investigation of physical phenomena, the lifeblood of physical geography, it is a surprise that it is not considered in more detail in the appropriate literature. Procedures of standardisation are often so entrenched within the traditional practices of subject that they pass without comment, are relegated to methodological footnotes or dismissed as irrelevant methodological niceties. How agreed and common procedures for measurement arose and the assumptions and often highly personalised battles that classes embody are lost in their routine application. The discussions, the compromises, the accepted exceptions and the basis for redefinition of the standard are not part of the usual publication diet of physical geography. The sociological and therefore by implication subjective and relativistic connotations are not seen as part of an objective study of physical phenomena.

Case Study

Classification of soils

Soil classification is essential to help clarify a complex phenomenon. Although most authors accept that classification is imperfect, some believe that certain types of soil classification are superior to others.

> If, however, the criteria of grouping are based on intrinsic properties not specifically linked to the objective [of the study or activity] it is hoped to achieve they may be called 'natural'. For the broadest use of 'natural' systems of classification are likely to be more generally helpful and therefore should be based on fundamental properties of the objects in a natural state.
>
> (Townsend, 1973, p. 110)

The quote illustrates that Townsend views intrinsic properties as the key determinant of a natural versus artificial classification. This is a common view within soil classification. Inherent properties are viewed as objective criteria, yet defining what is inherent is difficult. Properties derived from the parent material within which soils form could be viewed as inherent, but these alter as the soil develops. Assuming soils progress through predefined stages of change could provide a basis for defining inherent properties based on the stage of development. This argument could become circular, however, as the stage will be defined by the properties. Another common belief illustrated by the quote is that a 'natural' classification will be superior to any other and by implication form a sound, objective basis for the definition of other, more purpose-orientated classification schemes. The implication is that somehow the 'natural' scheme proposed as the ideal has no purpose!

Other authors, however, take a more pragmatic view of soil classification.

> any classification must have a purpose and with soils the aims can vary widely. For the most part, general systems developed for the definition of map units in soil surveys designed to aid land use will be discussed, but others have been constructed with less practical aims and more emphasis on soil formation and evolution to summarize relationships for scientific purposes.
>
> (Clayden, 1982, p. 59)

Classification is justified as long as it is related to a specific purpose and is consistent and coherent with that purpose. Whether this same scheme will be of use for other purposes is open to debate even if the properties that seem to be being used for classification are similar. This has important implications when classification schemes are used to produce products such as soil maps for agricultural productivity. If the classification schemes, irrespective of their purpose, are referring to the same phenomena in reality, then overlaps or commonalties between schemes may be expected. These arise not only because the theory underlying identification and description of the phenomena inform both schemes, but also because theory also informs the properties to be measured and their means of measurement. This means there may be convergence in the characteristics considered in classification schemes despite their different purposes.

Clayden (1982) makes the point that there are two potential ways of classifying soils. Hierarchical classification can be used. This framework has a 'family tree' structure with an initial 'type' defining membership of the class in terms of a range of properties and finer divisions being made within these properties to refine the classification. An example of this is the USA soil taxonomy developed by the Soil Conservation Service (now the Natural Resources Conservation Service) of the US Department of Agriculture. Soils in the hierarchical system are divided on the basis of precisely defined diagnostic horizons and soil moisture and temperature regimes. Soils are then classified into orders, suborders, great group, subgroup, family and series.

An alternative means of classification is the co-ordinate classification system. In this system entities are still differentiated, but the properties used are not necessarily arranged in a hierarchy or even ranked. The former classification type is, according to Clayton, easier to remember as properties are prioritised by importance. Although the latter classification scheme has been used in Russia, Germany and Belgium, the number of classes that are needed and their unstructured nature has meant these systems have not been taken up internationally.

Classification systems such as that employed by ISRIC (1994) are the result of a long process of negotiation between usually national agencies. The first FAO-UNESCO soil map of the world began in response to a recommendation in 1960, with the first maps being published in 1969. From that date, numerous meetings have been held to develop a common basis for classification, a common nomenclature and common methods of analysis. The result is not an agreed and single coherent framework based on a unifying theory or agreed hierarchy. Instead, the product is a compromise between existing systems developed for national and even regional interests and a need for international clarity in what researchers are referring to. The result is a classification scheme full of familiar local names such as chernozems and kastanozems, mixed with rigorous methods of demarcation. Clarity is provided by the use of two sets of properties for defining soils: diagnostic and phases. Diagnostic horizons are used for identifying soil units, and their presence is based upon the operation of specific diagnostic processes in the soil. Diagnostic properties are soil characteristics that do not produce distinct horizons but which are of importance in classification. Terms such as andic properties (exchange capacity dominated by amorphous material) and hydromorphic properties (water saturation conditions) reflect such properties. Surface diagnostic horizons are defined so that their main characteristic properties are not altered by short-term cultivation. This is an attempt to permit the application of diagnostic horizons to lightly cultivated areas. Phases are features of the land that are significant for its use and management. These features enable land use and its impact to be considered in the classification.

Significantly, it is the soil in its 'natural', uncultivated state that is seen as the basis for classification. Soil altered by cultivation is viewed as a secondary or inferior entity with secondary properties. This reflects a general attitude in soil classification schemes – 'natural' first then modifications to the natural. Natural relates back to the various theories embodied by the different classification schemes the FAO/UNESCO classification tries to accommodate. These theories reflect a concern with understanding the causes of soil formation, primarily the role of climate. Latitudinal and altitudinal variations in climate are matched by variation in soil properties and modes of development. Deviations from these 'model' forms are viewed as oddities within specific developmental sequences and are treated as such.

Events – mega entities?

Events, as noted above, are difficult to abstract from their context, despite this abstraction being essential for research aimed at establishing universals about the nature and behaviour of events. Whitehead (1929) views events as being totalities or holistic wholeness of everything at an instant in time. Events become the state of the universe at an instant and in that instant everything is connected to everything else. A specific event, or rather an instant, enfolds within itself all of space and time. Running instances together form a duration of events. Lacking such omnipresence, research into events is more practical if temporal and spatial boundaries are imposed to enclose an event. Such boundaries are defined by the purposes of the research being undertaken and, as such, events can be viewed as socially constructed but their existence is grounded in a reality knowable through scientific study.

Events, however, are not just a single entity, even if they are studied as such. Rather, an event is a group of entities, or an assemblage, related together within the temporal and spatial boundaries fixed by the researcher. An event could be viewed as a 'mega-entity' through the relational definition of its nature. Viewed in this way an event has the same philosophical and conceptual issues as individual entities with the added problem of its relational nature. As an event is a relational 'mega-entity' it could be analysed using relational concepts such as actor-network theory (Murdoch, 1998; Callon, 1999; Latour, 2005) or assemblage theory (DeLanda, 2006). Assemblage theory in particular, might provide a useful conceptual angle into the analysis of events as well as connections to a critical realist view of reality. Within this framework events are assemblages of entities, each entity containing within itself the potential for relations and actions or effects both realised and unrealised within a particular context. The event, a specific configuration of entities, provides that context within which entities are related to each other externally and by so doing define themselves through these interactions. Repeated configurations of the same entities, seemingly the same or similar events, provide the basis for identification of regularities, the appearance of universals to be abstracted from the particularities of a configuration. Repeated configurations trigger the same external relations between entities and so activate the same internal relations that make up that entity. A particular configuration does not necessarily provide insight into every aspect of the internal relations of an entity however. It is the external relations between entities that construct the specific entity we understand and know in a configuration. Alteration of these external relations, known or unknown to the researcher, could change the nature of a specific entity by triggering other aspects of its internal relations. These changes could involve, for example, the extension of the spatial and temporal boundaries of an event, the alteration of the entities involved in producing the researcher-defined event or a change in the wider context of the event brought about by environmental change.

The supposed 'universals' derived from studying events are always going to be conditional on context. The robustness of the external relations that determine the activation of specific properties of an entity will vary with both the context and the entity itself. This means that although we can explain events by reference to generalities, the specifics of their activation will make each event unique in its detail and so open to a range of possible interpretations and linkages to other events and explanations.

Summary

Entities are the centre of any study of reality. Entities are the units that we believe exist and which we can manipulate and use for testing ideas. Entities have properties or attributes associated with them, but not all of these are essential for defining what the essence of an entity is. Entities and kinds are intimately related. Kinds provide a template for what it is to be a member of that kind. They dictate the essential properties that define a member of that kind as well as the behaviour to be expected. The kind becomes the focus for explanation; the entity becomes an illustration of the kind. There are different types of kinds and it has been argued that research should focus on natural kinds. Even if geographic natural kinds exist, all kinds are defined for a purpose. This means that all kinds are to a large extent observer-defined and defined for a purpose, rather than defined as some natural building block that forms the 'real' basis for explanation. Kinds and entities are enmeshed within a network of signs and their meanings are likely to evolve as their purpose alters. Hacking (1999) identifies two types of kinds, indifferent and interactive. Physical geography deals with indifferent kinds, those that do not respond to the act of classification.

Chapter 4

Forms of explanation

Explanation in physical geography

This chapter outlines and justifies the different types of explanation that are regarded as acceptable within physical geography. Central to each mode of explanation are the researchers themselves. By defining the nature of the study, the questions to be asked and the entities to be studied, the researcher also sets the criteria by which a study is judged to be successful (see Chapter 10 for a discussion on the social context in which physical geographers carry out their research). Despite the wide variations in underlying philosophies, what is regarded as an acceptable explanatory framework needs to be established in order to understand where certain philosophies may be more appropriate for certain types of explanation and subject matter.

What is explanation?

The question what is explanation is a difficult one to answer despite the seeming simplicity. Although it is difficult to separate explanation from its philosophical framework, a general discussion of what explanation is may help to provide an outline of what physical geographers would find an acceptable explanation.

Explanation can be summarised as providing an answer to the question '*why?*', or, more specifically, '*why is something as it is?*' Given what we know about '*why*', what should we expect the something to do in the future? In answering these questions our explanations reduce uncertainty about reality and the future of our reality. Explanation and prediction are not the same thing, but they are related to each other.

Explanation could be viewed as a need to make the unexpected expected – to reduce the amount of surprise that anyone encounters. Surprise is replaced by expectation. When expectations are confounded by the unexpected, people tend to want to know why (Toulmin, 1960). The impetus for explanation may be that simple. Some arguments could be constructed about the evolutionary advantage such a search might provide, but such speculation is not really a concern here. An argument could also be made that replacing the unexpected by the expected reduces individual and social stress. By reducing unexpectedness, explanation could also be viewed as beginning to represent reality as it really is. Stating that something has an explanation is tantamount to stating that we have been able to identify what reality is and from that identification been able to understand how reality works. Explanation and understanding of reality come together. Explanation implies a mirror has been held up to reality and an undistorted, or at least acceptably distorted, reflection obtained. Whatever the case, the impetus immediately pushes the interaction between an individual and their surroundings to the fore. The need for explanation is based upon interaction, it is based in a need, an impulse to know about the environment and its behaviour.

The impetus for explanation need not be the same for different individuals. The example Toulmin uses is that of a stick appearing to bend in water (see Harvey, 1969). For some people this is a '*so what*?' moment. The significance of the interaction with reality is of no concern to them. For others, it poses questions about their reality. For the latter group of people, explanation provides an insight into how reality really is. Explanation produces a possibility of organising reality into simple categories – what is explained and what is not. Additionally, these categories can be refined into, for example, what explanation A explains, what it does not, what explanation B explains and what it does not, and so on. In other words a systematic body of explanations can be built up and used to define which parts of reality, or rather our interaction with reality, are explained and which are not. Viewed in this manner, explanation needs to be contextualised. Without some way of classifying reality, without a particular means of interacting with reality, the sources of unexpectedness that provide the basis for explanation are unknown. Training and experience in what to expect and what is unexpected, *and* the significance of each within one's frame of reference, needs to be identified. In this manner explanation, theory and philosophy are interwoven.

Accepting an explanation as valid implies that there is a set of criteria by which an explanation is judged. Once again the context of explanation becomes important. What is thought to be a valid explanation tends to depend upon the rules and conventions of a particular group. This could be interpreted as saying that explanation is socially determined. The exact rules may depend on social factors, but it could be argued that the need to predict expectations of reality from explanations means that the rules must converge to provide a version of reality adequate for the purposes of that group. This could be taken to mean that rules for an acceptable explanation should be rules that permit explanation to reflect reality as it is. This would severely limit the nature of the rules that could be applied.

Hempel (1965) suggested that scientific explanation was concerned with both universal and statistical laws. Universal laws are general principles, possibly derived from observation of regularities in reality. Given an initial set of premises, universal laws are applied to these to logically deduce an outcome. This is the classic covering law model. Universal laws usually imply that we have an idea of the reasons why premises and outcomes are linked, we have some mechanism in mind. This mechanism will always produce the outcome given the premise. Statistical laws can be applied in the same manner to premises and outcomes; the covering law model can be used, but we may not be certain that we will always get the outcome. The laws may be statistical because we do not understand the mechanisms involved or because they reflect observations of common conjunctions of premises and outcomes. Statistical laws, unlike universal laws, are only true because we observe them with our current level of knowledge; unlike universal laws, they have the potential to change (Salmon, 1998).

Causality

An important feature of any explanation is the idea of causality. At its heart the principle of causality assumes that every action, every event observed, has a cause. The event is determined by the cause. An event is as it is because it has been determined by events before it in a long chain leading back to what a researcher would identify as the cause. This means that physical geography and physical geographers tend to believe that events they observe are determined, that there is some knowable explanation, some knowable cause (or causes).

The simplest view of causality is the idea of cause and effect. An event happens, it happens prior to and probably close to another event. The first event can be interpreted as the cause of the second event – the effect. Cause and effect imply temporal and spatial proximity, although their linking may also require some ideas of why they are linked – a theory.

Simplistically, there are two views of causality, the successionist and the generative. In the successionist a cause is merely something, an event, which happens before something else. Connecting one event to another is something humans do, but the connections are merely our constructions and they have no existence in reality. In this view of cause, the entities that form the events that occur in succession are passive objects. They do nothing to influence each other and they do not interact in any manner to produce events. The generative version of causality takes the opposite view. In this view an event is generated by a cause and so the two events are linked and dependent upon each other. Taken together, cause and effect could be viewed as forming a single event – they cannot be considered separately as single events. A specific cause must generate a specific effect of necessity (Harre, 1985). In this view it is the nature of the entities themselves that generate the relationship. The nature of the entities involved in an event determine that event. An entity, by its very nature, being of a particular type, must possess particular properties that determine the manner in which it must, of necessity, behave under a given set of circumstances. Abstracting causality from the entities involved can produce information about general relationships, but it also tends to highlight the successional view of causality as cause is displaced from its context.

The choice between these two views of causality has been the core of much metaphysical debate (e.g. Cooke and Campbell, 1979; Holland, 1986; Oldroyd, 1986), but the debate largely boils down to issues considered in previous chapters on the nature of reality and human ability to understand reality as it is. Hume viewed cause and effect as a succession of events linked together (by humans) by an assumed mechanism. This was a human construction based on experience and not a real structure of reality. Rescher (1991, 1992) views causality as part of our conceptual framework for bringing shape and order to a formless reality. Much like Hume, the researcher is the centre of the construction of causality.

A generative view of causality sees events as linked by necessity and therefore reflecting reality as it is. Entities possess within them the potential, depending upon circumstances, for causing effects. Causality is about real relationships between entities and their actions. The key problem is that humans have no absolute knowledge of reality, no privileged position from which to view things as they really are. Kant tried to resolve the two views by invoking two realms of existence, the noumenal and the phenomenal worlds. The noumenal world is reality as it really is. Humans do not have direct access to this world, instead all they can know is the phenomenal world, the world of phenomenon as constructed by humans. Causality then becomes a messy mix of the two. Cause and effect(s) are generated by entities and their relations, but all we as humans can know are the succession of events. It is this succession of events upon which we impose an interpretation based on what we believe to be the 'real' relations between the 'real' entities based on 'real' laws. All the 'reals' are human constructs, but constructs based upon a participation in reality, an active structuring of phenomena.

Bunge (1962) identified three meanings of causality. First, causation which associates a particular event with a particular result (cause-effect). Second, the causal principle, which casts laws in terms of cause and effect. Third, causal determinism; a doctrine that asserts the universal validity of the causal principle. The three meanings are not unconnected. Assuming that everything is explainable by cause and effect relationships implies acceptance of a causal deterministic view of reality. It implies that cause and effect are observable because of the underlying importance of laws in determining how reality works. Within science, causality has a more restricted scope and definition. Humean cause is based on constant conjunction (Peters, 1991). Cause is always accompanied by an effect, with the cause occurring before the effect. This conjunction was believed to be sufficient to provide a basis for establishing a causal link. Such constancy in conjunction is difficult, if not impossible, to achieve in a field science, so more rigorous concepts are required. The operation of laws ensures that reality is composed of events that are causally linked both spatially and temporally – cause and effect relationships.

The presence of cause and effect relationships can be interpreted in terms of necessary and sufficient conditions. Any event could be thought of as having a set of causes associated with it. These causes are the conditions required for an event to take place. A potentially infinite number of conditions are not capable of study, so a subset(s) of conditions are the focus of study. These are the necessary conditions and the sufficient conditions. Necessary conditions refer to the conditions that would justify the non-occurrence of an event, whilst sufficient conditions are the conditions that would justify the occurrence of an event. Necessary conditions are more restrictive than sufficient conditions. Necessary conditions define the set of requirements that must be present if a certain event is to happen. These conditions can only be identified, however, if the event is prevented from happening by these conditions not being present. They imply an ability to manipulate reality to prevent the necessary conditions in order to allow their identification, a luxury that is not usually possible in field sciences.

Case Study

Necessary and sufficient conditions

Simms (2002) identified a series of karst forms around the shores of Irish lakes that had a very specific distribution. The rohrenkarren (or tube karren) are vertical, upward-tapering, closed tubes (Figure 4.1). In a detailed study of the distribution of these

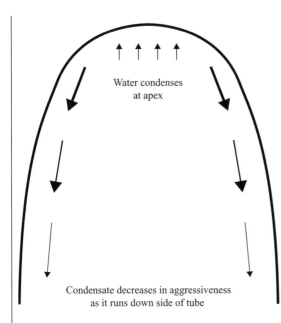

Water condenses
at apex

Condensate decreases in aggressiveness
as it runs down side of tube

Figure 4.1 Formation of tube karren.

forms, Simms found that the three lakes he used in his study, Lough Carra, Lough Corrib and Lough Mask, all shared a common feature – all were in a near permanent state of carbonate saturation. Active precipitation of carbonate was observed at all three sites. Karren tube formations were absent from lakes that were significantly undersaturated in carbonate. This means that carbonate saturation is a necessary condition for tube karren formation. The absence of this state means that this karst form will not be produced whatever the other favourable conditions for its formation. Identification of this necessary condition relied upon knowledge of the saturated state of other similar lakes, such as the Killarney lakes of County Kerry. Saturation was not itself, however, a sufficient condition for the formation of tube karren. There needs to be an additional set of sufficient conditions before tube karren can be formed.

Simms used his observations of the distribution of tube karren to advance a possible mechanism for their formation. Tube karren were confined to the epiphreatic zone produced by seasonal lake fluctuations. The level or height of the epiphreatic zone varied between lakes from 1 m to 3 m in tandem with the size or dimensions of tube karren found. Tube karren were formed by the condensation of water vapour in the air-filled apex of the tube (Figure 4.1). This water vapour is not saturated in carbonates, unlike the lake water from which it is derived. Limestone dissolution by the condensed water occurs at the apex of the tube. However, dissolution rates necessarily diminish as the condensed water runs down the tube. This means that the rate of expansion in the basal circumference of the tube is lower than at the apex, where dissolution is more effective. The almost perfect circular cross-section of the tubes reflects the uniform condensation of water onto the side of the tubes and the uniform dissolution that results. Enhanced rates of carbonate precipitation occur when the condensed water reaches the lake water. The release of carbon dioxide as a result of carbonate precipitation will enrich the atmosphere in the tube apex. More carbon dioxide is able to dissolve into the condensing water film and thus increase the potential for limestone dissolution.

Associated with these sufficient conditions is a necessary condition, the absence of which would mean that condensation and subsequent dissolution could not occur. For condensation to occur there needs to be a temperature difference between the rock and the water vapour. If both the rock and the water vapour were at the same temperature then condensation would not occur. The temperature of the limestone fluctuates at a higher frequency and amplitude than the lake water. This results in the two systems being in disequilibrium, at least for a few hours, when the rock is at a lower temperature than the water.

Simms also suggests that the initiation of tube karren may require a set of necessary conditions. Calm water is essential for trapped bubbles to initiate the indentations that can develop into tube karren. For such indentations to survive the lake water must be at the point of carbonate saturation. Only under these conditions can the minor irregularities persist to form tubes. The Killarney lakes of County Kerry, for example, had sufficient seasonal fluctuations of lake levels to permit the above mechanism to operate, but all were undersaturated in carbonate and so the initiation of tube karren could not occur.

Cause and effect relationships reveal the laws that govern reality. Underlying any cause and effect relationship is a set of structures. The operation of these structures ensures the consistency of the causes and the effects being observed. Where cause-effect relationships are not observed, these underlying structures either are not present or their operation is influenced by other structures. Identification of the underlying structures can be useful in identifying expected causal relationships.

Causal relationships can be represented by linkages between causes and effects. At the simplest level is the single cause and effect relationship (Figure 4.2). In this relationship, cause precedes effect and the link is assumed to be the causal mechanism. The link itself, however, could be viewed as being composed of a series of smaller events that link the cause to the effect (Figure 4.3). Linking a rainfall event to a landslide occurring may seem like a simple cause and effect. The link between the rainfall event and the landslide could, however, be viewed as being mediated through smaller and shorter duration events, such as a rise in pore-water pressure. When this series of links are observed for a number of particular instances of the phenomenon called a landslide, then the causal

Figure 4.2 Simple cause-and-effect relationship. The arrow indicates the direction of causality, from cause (C) to effect (E).

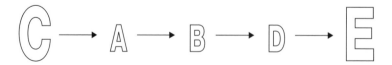

Figure 4.3 Cause (C) is connected to effect (E) by a number of intermediate events (A, B, D). The key question is whether A, B and D are always needed to produce E from C. If any one of the intermediate steps is not necessary, or can be substituted by another step, then the causal structure of C to E is open to change and possible instability.

structure obtains some sort of stability (at least in the mind of the researcher). A problem is, however, that the causal chain as perceived is capable of further refinement, of infinite regression reducing causation down to the lowest level.

The causal chain may be assumed to be linear and restricted, i.e. a single causal pathway. This is unlikely in reality. It is more likely that a single event at the start of the chain will propagate a number of subsequent events. In other words a single cause need not generate a single event, and nor need the links between a single cause follow a singular path to a particular event. There may be intervening factors that produce multiple potential pathways to an event or even series of events (Figure 4.4). A rainfall event may generate pore-water pressure sufficient to cause a mass of regolith to move, but it may also provide a lubricated layer over which the shear plane can move. In addition, a rainfall event may weather material in the mass of regolith, thereby weakening the coherence of the mass and increasing its susceptibility to movement. The single causal event can have a multiplicity of effects, all of which can contribute to the landslide, possibly even at a range of different temporal and spatial scales. Although the theory linking pore water pressure and rainfall may be all that is required to explain the occurrence of a landslide, the other effects are no less real and still contribute to the timing of the landslide.

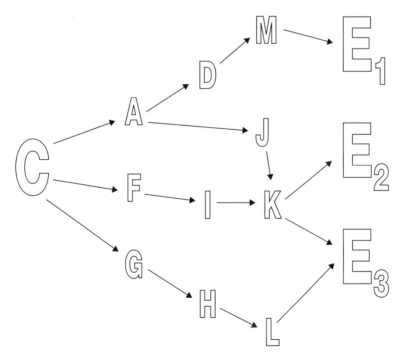

Figure 4.4 Illustration of multiple pathways to multiple effects. Note the impact of J in the network. J can be activated by A and so effect K. K can produce either E2 or E3, depending on its behaviour and specifically the behaviour of J, which in turn depends on A. This illustrates that once intermediate factors are viewed as part of an interrelated mesh of effects, the idea of a simple cause-and-effect relationship disappears. Instead, activation of different possible causal pathways becomes a concern.

A possible alternative view of cause and effect relationships is to view these as a causal web. Causes produce a range of effects that in turn function as a range of causes for other effects. Very soon a complex network of relationships can be built up between events with an associated range of outcomes. Tracing which of these relationships is necessary and which are only sufficient for any particular outcome becomes increasingly difficult as the complexity of the web grows.

Explanatory frameworks

Given that explanation is the goal of investigation in physical geography we need to look at how causal relations are put into an explanatory framework. Chapter 2 outlined the basics of deductive argument and this style of argument is useful for some causal webs. The deductive argument provides a means of linking a cause to an effect through the intermediate stages of effects. Each stage could be seen as a logical outcome of the previous stage and so form a string of logical conclusions and premises that lead from cause to effect. Clearly, the problem of infinite regress of cause has to be addressed at some point, i.e. at what level of reduction do you draw the line for an adequate explanation? Despite this problem, deductive reasoning provides a valid and formal basis for explanation. Deductive reasoning provides a basis for identifying what must be the case logically. The logical basis of the links is provided by reference to a 'law'. The law provides the reason why the cause and effect should be linked. It provides the logical basis for the effect. If C, the cause, happens, then the law derived from an overarching theory predicts that E, the effect, must happen. Put another way, if a state of reality exists at time t_0, which is knowable to the researcher, then applying law L, derived from a specific theory, predicts that at time t_1 there should be another state of reality. An outcome state can be predicted from the initial state if a law is applied to that initial state. There is a problem, however, with arguing, or rather presenting, cause and effect in this manner. Using 'state of reality' or 'state of affairs' as shorthand for the situation before and after a law has operated is very unspecific. Cause seems to be located within the phrase 'state of affairs' rather than in an object or entity. The phrase implies that had the state of affairs been different then the outcome may have been different. Questions of how different, and therefore what the role of any individual entity has within this explanatory framework, are not addressed by the phrase 'state of affairs'.

Induction, according to von Englehardt and Zimmerman (1988), refers to the situation where the controlling, or initial state of affairs, is known as is the resulting state of affairs. What is unknown is the link between them, the theory and a derived law that can connect the two states. Rather than arguing for a link between two states logically, induction *infers* the link between the states; there is no logical necessity for the inference to be true. All the problems of induction outlined in Chapter 2 are still relevant to this definition. If causation is located within entities and relations, then inductive reasoning is arguing that if cause and effect are known then the link can be derived. The manner in which it is derived is unclear and how the derived link can be assessed is also unclear. Any link considered as a likely candidate can, presumably, be argued for within some context, or else it would not be considered in the first place. If there were no way to obtain any other information about cause and effect other than that contained in the initial statement of positions, then there would be no way in which to compare candidate links. It is only the possibility of other means of assessment, or using other information that provides a way of assessing candidate links. Once the researcher moves into this form of analysis, however, it could be argued that they are moving from pure inductive reasoning and into a fuzzier arena of investigation and explanation.

There is a third style of explanation. Often the effects are known and there is a link – a law that produced them. The unknown is the initial state of affairs – the cause. The argument works back from effect to cause, from resulting state of affairs to initial state of affairs. Unlike induction, discussed above, there is a possibility of applying logical argument to the analysis. The law can be applied to the effect and a cause predicted in retrospect, in other words retrodiction can be used as the basis of argument. Given an outcome produced by a given process, we should be able to infer what the initial situation was. This style of argument is called abduction. A key limitation of abduction is that there is no logical necessity for the initial state of affairs to have existed. Unlike deduction, the logical final statement is not a necessary one. Only if the law assumed to link cause and effect (initial and final state of affairs) really operated can the argument be logically watertight. The initial state of affairs is the unknown, so there can never be certainty that the law operated, only a likelihood. There could be other laws that, when applied, suggest other initial states of affairs. There needs to be a way to assess which of the possible links back to a range of possible initial states is the most likely. In other words, abduction requires additional information in order to help in selecting what is viewed as an appropriate explanation.

From the above discussion, it should be clear that only the deductive argument produces a link between cause and effect that arises out of logical necessity. There can be no other effect, given the cause and the law operating. Admittedly, unless the deductive argument is somehow linked to reality, unreal statements can be logically deduced from inappropriate or unrealistic initial conditions. This means that much like induction and abduction, deduction requires information from outside of the confines of the logical structure of the argument to assess it. Deciding where, in the argument structure, to derive that information and how that information is actually of use in assessing different explanations, is a relatively little discussed area in physical geography. The importance of sedimentary profiles for reconstructing past environments is of vital importance in Quaternary studies, but why this type of information is used and how it intervenes in the explanatory structure is not usually made explicit. Datable sedimentary markers such as molluscs are vital because they provide a temporal framework for defining when events occurred. Likewise, pollen from sediment can be used to infer the nature of past environments, and how they changed over time. Event specific markers such as volcanic ash could be used to infer whether a specific event happened, and when it happened. When combined, this type of information can be used to test different arguments about the occurrence of specific events at specific locations, and at specific times. The type of evidence available may, however, focus an investigator solely on the type of events that can be identified and so necessarily limit explanations to these events alone.

In order to try to understand the importance of explanatory, or rather causal structures, in explanation in physical geography it might be useful to rethink the above discussions of different types of explanation as causal networks.

Deduction could be represented by Figure 4.5. Cause and law are known and the effect is a logical and necessary outcome of the two. In induction, only the cause and effect are known (Figure 4.6). A link could be made between the two, but its status is indeterminate and the number of possible links is high, if not infinite. In abduction, the effect and law are known, but the link is not stable nor is the status of cause fixed (Figure 4.7). Figures 4.5–4.7 highlight the stability and certainty of cause, effect, and links, or laws, between them. The situation is rarely as simple as illustrated in these three figures. Often there are a number of intervening states between the initial state and the resulting state. For deductive explanation, this is not a major problem as long as each link can be viewed as a logically necessary outcome from the previous set of conditions. The states are linked together by the operation of the law. As the number of intermediate states increases, it

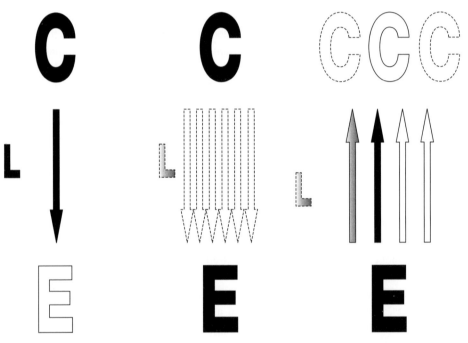

Figure 4.5 Structure of deductive argument, with known qualities shown in solid black lettering. Cause, or initial state of affairs, is known as is the law or process. Deduction logically derives effect, final state of affairs, from these first two. The arrow indicates the direction of the argument, predictive from cause and law to effect.

Figure 4.6 Induction. Cause (C) and effect (E) are known. What is unknown is how they are linked, if indeed they are. There are innumerable potential ways to link the two via a number of possible laws, but all links are untestable and so uncertain.

Figure 4.7 Abduction. Effect (E) is known, as supposedly is law (L) – shaded grey so conditionally accepted. Argument works from effect to cause. Law is applied and different plausible links back to cause made, but with differing degrees of confidence (different shadings). This means that there are different potential initial conditions (the 'dashed' Cs) that the effect could start from. Testing of the links back from the effect can be undertaken to select the most plausible scenario.

may become increasingly difficult to sustain the logical necessity of the argument. Uncertainty over what the states are, over the influence of other confounding factors and the possibility of 'other' outcomes can all affect the confidence a researcher may have in their explanatory structure. This may partly explain the concern with maintaining tight experimental control and keeping causal chains short, as this reduces the need to consider 'complicating' factors. Indeed, it could be argued that short and certain explanatory chains between initial, intermediate and final states are only possible under very specific conditions, and in relation to very specific questions. The movement for process-based geomorphology could be seen as an attempt to ensure that explanatory frameworks were based on short deductively argued chains. Although such studies might derive important general relationships between a specific set of initial conditions and a resulting state under the action of a single law, it is unclear what will happen to the nature of these relationships once they are moved beyond the confines of the controlled system.

Induction seems to have severe limitations as well (Figure 4.6). In the absence of other information, outside of the confines of the immediate statement of a relationship between cause and effect, the route between the two is almost completely indeterminate. There are a multitude of links that may be possible, that could be argued for, and there is no way of judging them. Similarly, a multitude of intermediate states can be suggested and a range of convoluted linkages put forward to enable even seemingly absurd explanations. As with deduction, maintaining a simple set of links may seem to make an argument more reasonable. Once there is the possibility of a range of possible links, the certainty of any single link is called into question.

Abduction operates as in Figure 4.7. The effect is known and, supposedly, so is the law (the link between the effect and cause). Working back along this chain, the cause can be found. Abduction is based on being able to tell a plausible story to link effect and cause together via a valid law. There are, however, potentially a range of laws that could be applied to explain any effect. By extension this means that there are a range of potential starting points, a range of potential causes by which laws operate to produce effects. This sounds much like the problem associated with induction – the problem of how to pin down the law and the cause. Unlike induction, however, abduction starts with both law and effect (explanation). In other words, it is assumed that the law is the correct one to start with. The issue then becomes forming appropriate links back to the cause. Linkages could be simple, such as a single one, or more convoluted such as requiring a large number of linkages and the operation of co-varying laws. At this point Goodman's (1958, 1967) assertion that simplicity in such reasoning is the important guide could be invoked to aid the selection of valid explanations. Put simply, Goodman stated, faced with alternative explanations, preference should be given to the simplest. Defining simplicity is not a simple task in itself however. Simplest does not just refer to how many links there are in a causal chain, but also to the ease of understanding, the relation of the explanation to other explanations of similar phenomena, and the context within which the explanation is to operate. A long causal chain or even causal web may be appropriate in some cases where a particular law provides a simple explanation in terms of always producing a given effect, but which requires a long series of events to link cause and effect. In other words, simplicity is not an absolute property of an explanation, but a contextual one.

Viewing abductive explanation as being contextually bounded is important. It highlights the conditional nature of this form of explanation.

Turner (2004) points out that two types of abductive argument exist. The first is for unobservable tiny entities, the second is for unobservable past entities. Unobservable tiny entities, such as electrons, are only known through their interaction with other phenomena in experimental situations. Their existence is deduced from their effects. Their use in explanation is that they function as a unifier for a range of phenomena that would be difficult to explain without them. Additionally, they can be manipulated to produce new phenomena through which new properties about them can be deduced. Unobservable past ('historical') entities cannot produce new phenomena as they cannot be manipulated. Historical entities instead serve as unifiers of phenomena only. Using analogues to contemporary events or entities may be the basis for believing an historical entity indicates that a particular event happened or that that entity behaved in a certain manner. Belief, for example, that a specific species of mollusc indicates a change from fresh to saltwater conditions may be based on observations and manipulation of current mollusc species. This does not negate the fact that the fossil mollusc cannot be directly observed or manipulated. The presence of the fossil mollusc in a sedimentary profile can, however, serve to indicate a particular environmental condition. The fossil mollusc acts as a unifying entity integrating within it information about the nature of a past environment, even if this information is derived by analogy with current environments.

Abductive argument can be bi-conditional or conditional. An abductive argument is classed as bi-conditional if the cause thought to result in the effect can produce only that effect. The usual statement of this is that if, and only if, C, then E is true. If, and only if, a specific cause C occurs then, and only then, will a specific effect, E, occur. There is, in other words, a necessity for a specific effect to follow only from a specific cause. In most cases, however, it is more likely that the abductive argument is only conditional. The link between effect and cause implies only a probable inference. The usual statement is along the lines: if C is true then E is true. In other words, a specific cause, C, occurs and a specific effect, E, occurs, but C need not always result in E and E need not only result from C. There is a varying degree of doubt that C will produce E. Abduction searches for the range of hypothetical conditions that could link effect to cause via suitable laws. The laws should link the premises, the cause and effect, in a manner that ideally produces a bi-conditional argument. Failing this, the conditional argument should be a highly probable or stable one (Inkpen and Wilson, 2009). At its heart, abduction requires a researcher who can select: selection of both what makes a suitable law – selection of causal pathways linking cause and effect – as well as selection of what is the 'simplest'.

Abductive arguments cannot be decided without recourse to reality, without interaction to determine which of the potential pathways linking cause and effect are present and which, if any, have or are operating. In reality, as von Englehardt and Zimmerman (1988) noted, science is about the interplay of all three types of explanation: deduction, induction and abduction.

Looking at Figures 4.5 and 4.7 it is clear that deduction and abduction are, in a limited sense, mirror images of each other. Deduction works from cause to effect, abduction from effect to cause. They work through the diagrams in opposite directions. The key difference is that deduction has a logical necessity about its outcomes, whilst abduction can only infer. Deduction permits only one, certain link between cause and effect, whilst abduction admits to an infinity of links, but each link has a different degree of certainty or probability. Inserting deductive arguments into an abductive explanation could provide the basis for limiting the potential pathways linking cause and effect, particularly if induction, in the hypothesis testing sense outlined by von Englehardt and Zimmerman, can assess the existence of the logical outcome in reality.

Case Study

Abduction as a form of explanation in environmental reconstruction

Zong *et al.* (2006) investigated changes in the intensity of the East Asian Monsoon system (EAM) during the Holocene by reconstructing freshwater flux variability in the Pearl River Estuary, Southern China. As with much research in environmental reconstruction, the behaviour of a larger-scale event, (in this case the EAM), is understood through the behaviour of a smaller-scale event, (in this case freshwater flux). There are a myriad of linkages assumed between the larger-scale event and the smaller-scale one. Proxies (indirect indicators of former climates or environments) are directly related to the smaller-scale event. The value of proxies in identifying the behaviour of the larger-scale event depends on the accuracy of the relations between the events at different scales. These scale relations need to be considered and assessed as well as the relations between the proxies and the smaller-scale event. In this example, the 'effect' is the proxy data (fossil diatom assemblages, bulk organic $\delta^{13}C$ and C/N values) and the inferred 'cause' is changes in EAM strength (a large-scale event), which is transmitted through a smaller-scale event (changes in freshwater flux). Zong *et al.* (2006) satisfied the following conditions allowing a link to be made between the effect and the cause: (1) Changes in the freshwater flux of the Pearl River Estuary are strongly linked with EAM variability; (2) the proxies used respond to, and faithfully record, changes in freshwater flux over time; (3) the combined evidence favoured one of a number of causes (represented as 'events' in Figure 4.8). An *et al.* (1993) demonstrated that variations in the intensity of the EAM result in fluctuating precipitation and freshwater flux in the Pearl River Estuary. Diatom assemblages change systematically along the axis of estuaries in response to a change in the salinity gradient (e.g. Amspoker and McIntire, 1978; Zong *et al.*, 2006), whereas bulk organic and suspended sediment $\delta^{13}C$ and C/N change systematically along the axis of an estuary reflecting the change in organic matter source (e.g. Middelburg and Nieuwenhuize, 1998; Wilson *et al.*, 2005; Zong *et al.*, 2006). The authors use the contemporary spatial relationship between proximity to a freshwater source (used here as a surrogate for freshwater flux) and environmental proxies to interpret changes in freshwater flux to the Pearl River Estuary in the Holocene. The fossil diatom assemblages indicate a reduction in water salinity, particularly between 7,500 and 6,000 cal. yr BP. This coincides with an increase in organic matter derived from freshwater sources, as indicated by a fall in $\delta^{13}C$ and a rise in C/N. Because changes in the diatom assemblages and in the bulk

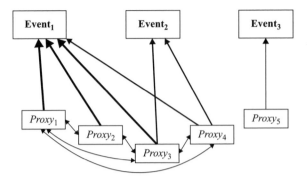

Figure 4.8 Proxies (the 'effect') have many possible relations linking each to different events (the 'cause') with differing degrees of belief or certainty (represented by the thickness of the arrows – thicker arrows implying increased certainty). Proxies are also linked to each other in the sense that contemporary studies find such proxies occurring together. The occurrence of specific proxy ensembles may be the necessary outcome of the operation of contemporary processes and so by implication identify the same processes as having occurred. The figure suggests that Event 1 is the common cause. Event 2 is possible as a common cause but with less belief or certainty as less proxies are related to it and with a lower degree of certainty for all proxies. Event 3 is only related to Proxy 5 which itself is an unexpected proxy as it is unrelated to the other four proxies. Reprinted with permission from Inkpen, R. and Wilson, G.P., 2009. Explaining the past: abductive and Bayesian reasoning. *The Holocene* 19, 329–334. © SAGE Publications.

organic $\delta^{13}C$ and C/N occur independently of each other, but are driven by a common forcing mechanism – proximity to a freshwater source – the authors conclude that an increase in freshwater flux has occurred in the Pearl River Estuary and they interpret this as indicative of an increase in the strength of the EAM at this time (Zong *et al.*, 2006).

It is useful to consider the interpretation of the data by Zong *et al.* (2006) and their selection of a causal pathway. Zong *et al.* (2006) test the hypothesis that 'palaeo-environmental proxies from an estuarine environment can be closely related to fluctuations in freshwater discharge associated with monsoon variability' (Zong *et al.*, 2006: 252). As shown in Figure 4.9, however, more than one cause (or 'event') can often explain the effect (proxy evidence). In this example, an increase in freshwater flux and a decrease in relative sea level are equally probable based on the nature of the proxy evidence alone, and neither of these events are mutually exclusive. For example, salinity and organic matter source may change in response to a fall in relative sea level (decreasing marine influence at the core site resulting in an apparent 'increase' in freshwater flux). The two events (increasing freshwater flux and falling relative sea level) are equally probable based on the available evidence. However, acceptance of either event will imply the operation of related events at a larger scale (the scale of interest in this study). Acceptance of a fall in relative sea level as the probable event necessarily requires the operation of larger-scale processes (e.g. isostatic changes, 'eustatic' change, etc.). Because relative sea level was actually increasing during the period in question (Zong, 2004), an increase in freshwater flux as a result of a strength-ened EAM remains the most probable event that can account for the evidence.

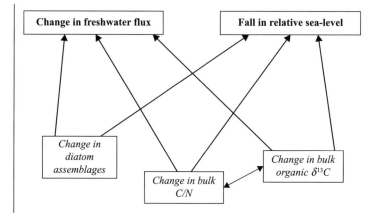

Figure 4.9 The evidence (change in the multiproxies) can be fully explained by each event (increase in freshwater flux and a fall in relative sea level), which are not mutually exclusive. However, knowledge of larger-scale events, in this case a rise in relative sea level, leaves an increase in freshwater flux as the most probable event in explaining the evidence. Reprinted with permission from Inkpen, R. and Wilson, G.P., 2009. Explaining the past: abductive and Bayesian reasoning. *The Holocene* 19, 329–334. © SAGE Publications.

Consideration of the proxy data within the context of larger-scale processes, therefore, alters the likelihood of specific events accounting for the proxy data.

Any study of causality in science could involve developing an abductive framework within which competing or different potential pathways can be subjected to deductive arguments and empirical testing. The exact make-up of any explanatory framework, whether it is 50 per cent deductive or more, is largely irrelevant. What is relevant, as critical rationalists have noted, is that any explanatory structure is capable of being tested at some point within its structure. The possibility of failure of a structure, the potential for falsification, remains a central feature of scientific argument, but it needs to be placed within a larger context of the logic of the intermingling explanatory frameworks of which it is a part.

Accepting the above argument that science, or more specifically a field science such as physical geography, is about constructing explanatory frameworks that are both abductive and deductive, the next question is how can we test these frameworks? Figure 4.10 illustrates a possible set of cause and effect relationships and is based on diagrams from Pearl (2000), who also provides a probability based justification for this approach. In Figure 4.10a cause and effect are clearly linked, one follows of necessity from the other. In the next case, in Figure 4.10b, the situation is complicated as there is now a third factor, Z, which could cause both C and E, breaking the causal link between C and E. C is reduced to an outcome of a common cause of Z rather than the cause of E.

How can we decide which one of the causal frameworks is correct, or rather which is more appropriate for our needs? In the cases above there is no way of deciding which causal framework is the most appropriate. Any information can be interpreted in favour of any framework. The issue can be resolved if there is an intermediate effect in the graph,

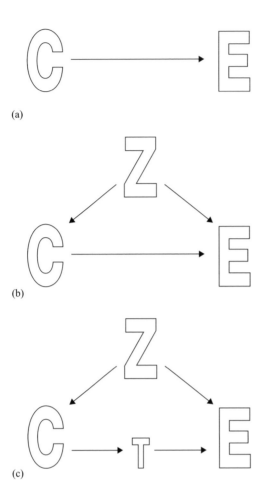

Figure 4.10 Initial simple cause and effect relationship is complicated by a confounding factor (Z). Introduction of an intermediate factor (T) between C and E provides an opportunity to assess cause-effect relationship. Eliminating pathway C to T means that only Z can cause 'E'. If E does not occur in this situation then it implies that C is required for E. Eliminating pathway T to E, whilst retaining Z, has the same result. This highlights that causality is determined as much by the activity of doing, of finding appropriate intermediates to eliminate or control for, as it is by theorising.

in this case T. What T provides is the possibility of constructing a set of relationships in which the causal linkage of one factor can be eliminated. Figure 4.10c outlines how this can be done in this case. The link between C and T and between T and E can be removed. The figures illustrate the effect of controlling the influence of one factor whilst letting the other factor vary. This is a simple illustration of what researchers have done for years in experimental designs. Assessing the influence of one factor whilst keeping others constant is not a new idea, but within the context of causation, it highlights the importance of understanding cause by intervention. Causation is no longer seen as something that can be derived from theory alone. Determining causation, deciding between causal pathways, requires intervention. Causal networks highlight that causation is a product of human intervention in a system rather than a property existing independent of that intervention. Causation, in other words, has no meaning without human intervention.

Constructing causal networks also highlights another property of causation. A set of entities and relations can be identified and defined as causal networks of a general type. Any landslide, for example, will be expected to consist of a set of entities and relations.

The occurrence of a particular landslide will involve the activation of specific pathways within the causation network. The causal network is the general or type level, whilst the specific landslide is the token level (Pearl, 2000). The token level is an instance of the general or type level. The two levels are structurally the same; it is only the specifics of each landslide scenario that differ. The two levels interact and inform each other. More token-level events can help to clarify and refine the type-level structure, whilst the description and explanation of token-level events relies upon the structures developed at the type level. In this manner, the stability of certain causal networks is established; more token-level events confirm the structure of the type level. Once a researcher becomes confident of the nature of the type-level causal structures then prediction and extension of the explanation can take place. This may involve stripping the type-level event down to what the researcher regards as the essential components and modelling likely behaviour based on these. The constant interplay between type- and token-level structures can become ossified if the research community believes that it has clearly established the causal network. In this case, the measurement of an effect and its explanation may become so unproblematic that the effects are not questioned and form a building block in the development of explanations for other effects.

Case Study

Linking theory and practice

An illustration of linking theory and practice is provided by Price *et al.* (2012) in their analysis of the validity of the metabolic theory of ecology. The metabolic theory of ecology (MTE) is probably best known within physical geography as the relationship between body size, body temperature and metabolic rate in organisms. For most organisms their metabolic rate is scaled to the organism's mass to the power of ¾. This relationship holds not just for single organisms but across a range of scales integrating cellular- and global-scale process (West *et al.*, 1997). Despite its scope, the theory has not, in the view of Price *et al.*, been tested to assess the validity of its assumptions and the growing group of extensions that have attached themselves to the central theory.

Price *et al.* (2012) identify four levels of the theory that they can assess: level 1, the internal consistency of the underlying theory; level 2, the validity of the assumptions of the theory; level 3, the predictions of the theory; level 4, the extended predictions of the theory. For level 1 they suggest that there is a valid basis for the theory in that the basis of the theory can be derived from the assumption that mammals have evolved a blood vessel network that minimises energy loss through dissipation and wave reflections, a Lagrange optimisation scheme. This ensures the internal consistency of the theory but also, importantly, immediately provides a means of translating the theory into an empirically measurable set of entities, entities related to the blood vessel network and networks in general. They state that currently there is no known way to construct a Lagrange optimisation scheme for minimising energy loss for pulsatile flow resistance for whole networks (Dodds *et al.*, 2001; Apol *et al.*, 2008). This means that the empirical statements have been derived as best approximations

> from a theory that is not capable of exact translation. This implies that there could be other translations that emerge as the theoretical basis of optimisation becomes clearer.
>
> For level 2, the assumptions are expressed in terms that are translated into empirically testable statements, so linking theory and empirical reality. The empirical information, they contend, involves determining the dimensions and properties of physical networks and the rates of flow and oxygen exchange, but they accept that in some cases, such as the measurement of all capillaries, empirical testing is not practical. By comparing the assumptions to empirical data they conclude that most of these assumptions are met in real vascular networks but that mammalian vascular networks are not consistent with the 'strict' assumptions of the theory, although in some cases the available data are very limited. This begs the question how far from strict adherence do the empirical data need to be before the theory is unacceptable? At level 3, the empirical data suggest that the relationship does hold but not across the whole range of scales initially suggested. The relationship seems to be a useful generalisation for larger organisms.

Another explanatory framework is Inference to Best Explanation (IBE). This is a form of abductive reasoning (Harman, 1965), which consists of making inferences about potential causes of an event or observation. The difference to abduction (described previously) is that the researcher infers that one hypothesis is a 'better' explanation than another.

> In making this inference one infers, from the fact that a certain hypothesis would explain the evidence, to the truth of that hypothesis. In general, there will be several hypotheses which might explain the evidence, so one must be able to reject all such alternative hypotheses before one is warranted in making the inference. Thus one infers, from the premise that a given hypothesis would provide a 'better' explanation for the evidence than would any other hypothesis, to the conclusion that the given hypothesis is true.
>
> (Harman, 1965, p. 89)

This raises the key question about what makes one hypothesis a 'better' explanation than another hypothesis? If it is based on the experience of the researcher then this explicitly brings subjective aspects into hypothesis selection, and by extension judgement.

Lipton (2004) suggests that finding the 'best' explanation from the host of possible hypotheses requires the finding of the inductive inference that tells us on what basis we can judge each result as being more likely. In addition, Lipton is concerned with identifying the 'loveliest' explanation. As Walker (2012) notes, this has nothing to do with aesthetics but is concerned with the explanatory virtue of an explanation. Based on the principles used and their application, the 'loveliest' explanation provides both breadth and depth to the narrative derived to explain the evidence. So Lipton and Walker are concerned with the explanatory 'depth' or 'scope' of each explanation derived from each hypothesis.

Glass (2007) suggests that there are two components to 'best explanation': the account of what constitutes an explanation and a suitable methodology for comparing competing explanations. In a more recent paper (Glass, 2012) he states that there are three classes of approaches to identifying a 'best explanation': (1) Bayesian, (2) those based on confirmation theory, and (3) those based on coherence. The Bayesian approach assumes that the 'best explanation' will also be the one that has a higher posterior probability. This guarantees that the 'best explanation' is also the more probable one given the evidence but, as Glass notes, this trivialises IBE as it replaces 'best' with 'most probable' and doesn't address the explanatory depth and scope issues raised by Lipton and Walker.

Confirmation theory assesses the degree to which a piece of evidence confirms (or does not confirm) a hypothesis. Unlike Bayesian approaches, confirmation theory looks at the impact of evidence on the probability of the hypothesis and not just the posterior probability of the hypothesis given the evidence. A subtle distinction perhaps but one that focuses on how much more probable a hypothesis is given a piece of evidence. Coherence measures (also known as overlap coherence measure or OCM, Olsson, 2002) rank explanations based on the evidence for each. Glass (2012) suggests that this last approach provides a basis, based on simulations of outcomes of hypotheses, on which to compare competing hypotheses but does not define what constitutes an explanation. Rather, the method provides a means of identifying the degree of coherence between evidence and explanation.

Walker (2012) takes a different tack to identifying the 'loveliest best explanation'. He identifies two key issues with IBE: the subjective objection and the truth objection. The subjective objection states that 'loveliness' is too subjective a term to guide inference. Failure to converge on the 'loveliest explanation' would be confirmation of this issue. The truth objection states that just because an explanation is 'lovely' doesn't mean to say it is true as well.

Using Kuhnian ideas about research and the development of ideas, Walker suggests that both objections can be overcome. In answering the subjective objection, Walker suggests that scientists do tend to converge on a single explanation when confronted with competing hypotheses, as Kuhn noted. Exemplars of solving puzzles or problems are often used as templates and these provide scientists with 'standards' of 'loveliness' against which to assess potential explanations. This does, however, mean that 'loveliness' may be contextually dependent, in the sense of being specifically developed and assessed within a particular location and time. 'Loveliness' therefore could reflect localised (and disciplinary) standards.

By rejecting Kuhn's relativism for scientific ideas (a cornerstone of his work), Walker is able to argue that the truth objection can be overcome as each refinement of the 'best explanation' or 'loveliest explanation' brings us closer to the 'true' explanation. He suggests that the reliability of explanations can be viewed as evidence that successive exemplars and their associated explanations are approaching the truth.

The problems of causation

Causation has been presented as unproblematic in the previous discussion but there is a philosophical debate about the nature of causation that could be important for how physical geographers view reality. A 'standard' view of causation (if there is such a thing) is provided by Woodward (2000, 2003). Woodward proposes two ideas that are essential to causation. First, there is the idea that explanatory relationships are relationships that could, in principle, be manipulated to tell us how other variables would alter. This is the view of explanation by intervention or manipulation. Explanation, and by implication causation, is only possible if what you are trying to explain is open to intervention so that you can alter relationships and observe or perceive an effect. Second, explanation and causation imply invariance. An explanation or cause is invariant if it remains stable under an intervention. An intervention is an experimental manipulation that determines whether changing one variable changes another. If the intervention results in this change for a whole range of changing conditions then the relationship is deemed to be invariant. It would then be possible to identify a range of varying conditions under which this invariant relationship held and identify this as the domain of invariance. According to Woodward (2000), it is this invariance that is identified in searching for explanation and causation and it is this invariance that, under sufficiently wide-ranging conditions, is generally viewed as a 'law'.

A key aspect of this view of explanation and causation is the concept of modularity. Modularity refers to the lack of connectivity between relationships in a system. Although Woodward expresses modularity in terms of equations defining such relationships, it is also possible to express the concept in terms of relationships alone. A system is modular if, and only if, each relationship is invariant under some range of interventions. For each relationship, however, it is also possible to intervene in such a way that only that relationship is disrupted whilst the other relationships in the system remain the same. This view of systems is an essential aspect of an interventionist's view of reality. It implies that once we have established the relationships that define our version of reality, then it is relatively easy to determine causes as all we need to do is to develop means of intervening in each relationship. The disconnectivity between relationships will mean that we can isolate and examine each relationship alone and determine its invariant domain independently of the other relationships in the system.

Cartwright (1979, 2003) does not agree with this view of explanation and causation. She takes a more relational view of explanation and causation that stresses the significance of context for modularity. Cartwright (1999, 2007) and Cartwright and Efstathiou (2011) view the search for causation as having one of two perspectives – causes are either hunted for or they are used, but it is not possible to do both simultaneously. Woodward's account of causality lies firmly within the realm of how to use cause, as far as Cartwright is concerned, as it involves manipulation and intervention to discover causation.

Cartwright and Efstathiou (2011) suggest that any attempt to search for causation is faced with two related problems: unstable enablers and external validity. Unstable enablers refer to changes in enabling factors that result in changes in the relationships that are thought to be stable and so causal. As contexts change the underlying structural arrangements of a system can alter and so the factors previously in a relationship that caused or produced the effect may no longer produce that effect. External validity refers to the problem of generalising from a particular context to another. Changing context changes the relationships. As causal relationships tend to be identified and defined within very narrow limits of experimental work, or with a specific population or data set, then changing these could alter the validity of the relationships identified. Extension of such results is essential for the transferability of causes and explanation, but each move away from the context (or even methods) of discovery provides the potential for the problem of external validity and unstable enablers. Cartwright and Efstathiou (2011) illustrate this issue with what they regard as a standard method for hunting causes, a Galilean experiment. In Galilean experiments the researcher eliminates all confounding factors so that it is then possible to establish law-like regularity between cause and effect in the isolated system. The physical removal of all confounding factors that interfere with their operation identifies cause and effect. This is the type of relationship that is used for reliable prediction. The problem is that there is nothing to say that this cause and effect relationship will persist once the confounding factors are present.

Cartwright and Efstathiou (2011) suggest that current theoretical accounts of causation based on hunting for causes do so by using different methodologies that enable cause and effect to be identified and labelled but only within the particular circumstances created by those methods. Once these circumstances change there is no sure way of knowing that the methodologies used before will work in the hunt and identification of cause and effect.

Cartwright (1983, 1999) and Cartwright and Efstathiou (2011) suggest that causation may be more appropriately considered by analysing capacities and dispositions. Capacities are powers that an entity or relationship possesses that contribute in a fixed way to an effect whenever they are present. We know about the capacity of an entity or a relationship from the influence it exerts on our observations, in the effects we measure. We do not necessarily know that that capacity exists independently of the circumstances in which the contribution is produced.

Summary

Explanation is a means of reducing uncertainty about reality. Explanation implies that there is an interaction between the researcher and reality, based upon an impulse to know about the physical environment and how it behaves. Deciding what is and what is not an explanation implies that there is a set of criteria for judgement. This suggests that explanation is a communal activity based on rules and conventions. At the heart of explanation is a belief in a causal relationship being present between entities in reality. Deriving where causality lies, in the succession of events or in the entities, is an important consideration. If causality resides in entities then they take an active role in determining the nature of reality.

Linking cause and effect is a difficult process and relies upon some underlying theory about how reality operates. This provides the basis for deciding that event A should be connected to event B; indeed event A should be the cause of effect B. From this simple relationship different explanatory frameworks can be imposed. Deduction assumes that cause and the connecting law or mechanism is known and the effect can be logically derived from these. Induction assumes that cause and effect are known but that the law or mechanism is unknown. Abduction assumes that effect and law or mechanism is known and a plausible cause can be inferred. Explanation usually involves a combination of all three explanatory frameworks. Deciding on the plausibility of a cause and effect relationship will involve the researcher intervening in reality to try to control for factors assumed to be important in the explanation.

Chapter 5

Probing reality

Probing and the dialogue with reality

Even when physical geographers have developed a theory about how reality operates they still need to determine if this theory 'works' in the 'field'. The 'field' refers to the arena within which the theory is assessed to see whether it operates as expected. This could be the field as understood as the natural environment, it could be the field as understood as a controlled subsection of the environment as in field trials. Laboratory-based and computer model simulations would also count as the 'field' in this context. 'Work' in this context is assessing whether the results of probing reality, its measurement, matches the expectations that have been derived from the theory.

The assumption is that the 'field' is the same reality whoever defines it, whoever studies it and whatever instruments or techniques are used to assess reality. Recognising that the field is defined in different ways is not just an acceptance that different physical geographers are doing different things but that they may be operating within and referring to different worlds. Despite these differences, physical geographers have to assume, as outlined in Chapter 2, that there is a physical reality behind their investigations. This assumption does not, however, mean that their 'fields' of practical research need to be defined in the same way, or approach each other in their characteristics. As noted by Goodman (1960, 1978), researchers are into 'world-making', into constructing their own worlds that they explore using techniques they have learnt. Important in such world-making are the purposes and objectives of the research. These begin to define the type of reality, the phenomena, which form the basis of the researchers' view of reality. These objectives define the type of study to be conducted, that is whether it is a laboratory-based experiment, a field-based experiment or a field study involving direct monitoring of the environment.

Any probing of reality is a form of intervention. Some researchers (Harrison and Dunham, 1998) suggest that, as in some interpretations of quantum mechanics, reality does not exist until the researcher intervenes. They suggest that reality does not cohere, the wave functions do not collapse until the researcher measures it. Interpreted in this extreme manner, this means that the state of a glacier, the velocity of a river, the entrainment of particles, does not exist or occur until the researcher measures it, until the researcher intervenes in the system. However, coherence of a physical system above the quantum level does not necessarily require a human observer. The quantum system is only observable under the unreal conditions of highly controlled laboratory conditions. Under these conditions the observer is the only entity that interacts with the quantum system. In the 'real' world, quantum particles interact and the wave function is resolved by these interactions with other entities (Spedding, 1999; Collier *et al.*, 1999). In other words, the uncertainties of the quantum level are resolved before the researcher intervenes and measures the system; the wave functions have been collapsed. This means that physical reality is not fuzzy and incoherent, nor in a series of potential multiple states before the researcher measures it. Reality is a resolved, singular whole. This does not mean that the researchers will have absolute knowledge of this singular whole, only that they will be interacting and having a dialogue with an entity that exists independent of them. This means that reality will operate and function as it does whether humans recognise it or not.

Intervention does, however, create different worlds, by creating different dialogues. The first step in this intervention is the framing of reality. Framing of reality involves deciding which variables are important, how these are to be defined and measured and, vitally, defining what is not important. This latter point highlights what is not to be defined and considered important to the objectives of the project. Undertaking laboratory experiments could be viewed as the most extreme form of intervention as the reality made becomes the experiment itself. A laboratory is used to exclude all reality other than the variables thought to be important for the operation of the part of reality under investigation. These variables are excluded because they are not thought to be important for the creation of the phenomenon that is the focus of study. Only those variables and relations thought, by some theory, to be important and involved in the creation of a phenomenon, will be used in a laboratory study. Even these variables and relations will be closely controlled and maintained at magnitudes and frequencies that can produce the phenomenon with the equipment available.

Laboratory studies require, therefore, a clear identification of the phenomenon to be investigated. This often involves transferring observations of changes in the 'field', the environment outside of the laboratory, and their interpretation as significant phenomenon. The laboratory conditions then try to recreate the phenomenon as observed. A key question that needs to be asked is whether the phenomenon being replicated in the laboratory is really the same phenomenon as that observed in the 'field'? Hacking (1983) suggests that it may be the case that the phenomena in experiments only exist in the experiments. The phenomena require the experimental conditions to exist. Rather than replicating the conditions in reality that produced the phenomenon, the experiment produces the only conditions under which a phenomenon can occur. These may be the only conditions in which this phenomenon can exist and there is no guarantee that these conditions exist in reality. In addition, there is the problem of equifinality in the production of phenomena in experiments. The experimenter can never be certain that the output from the experiment, the phenomenon, is only producible by the conditions in the experiment. There may be other conditions under which the phenomenon may be produced and these may be more reflective of the conditions under which it is produced in the 'field'. This problem of replicating reality, or rather the conditions that produce the phenomenon of interest, is a general problem in any study of physical reality. This problem was particularly acute within salt-weathering experiments in the early 1970s.

Initial experiments such as Evans (1970) and Goudie (1974) were designed to illustrate the potential effectiveness of salt as a weathering agent. The experiments were designed to maximise the destructive power of salt on rock. This meant that saturated solutions were used and salt supply was fairly continuous. This ensured a supply of an aggressive weathering agent to the rock. This is a situation unlikely to occur in reality. Within the confines of the experiment, however, this produced the desired effect – the rock broke down relatively rapidly. Given the extreme circumstances, however, it was possible the effect might only exist in the experiment itself rather than reflect the phenomenon thought to be observed in reality.

McGreevy (1982) makes the point that a great deal of the variation observed in the nature and rates of salt weathering in laboratory experiments may be the result of variations in experimental design rather than variations in the phenomenon itself. As the desire to make the experiments more 'realistic' increased, so did the potential variation in experimental designs. Alterations to variables such as salt concentration, salt supply, weathering conditions and sample shape all become potentially important factors in producing different degrees and types of weathering. The phenomenon identified in the 'field' became increasingly complicated and difficult to understand as attempts to model 'reality' in the laboratory became more 'realistic'. An important problem with these studies is the type of effect measured. The manner in which weathering was measured tended to be by weight loss or percentage weight loss of a sample. The relationship of this index of change to phenomenon observed in the field is unclear at best. Rocks undergoing salt weathering in the field exhibit specific forms of alteration such as flaking, pitting and crumbling. Loss of material does occur, but visually it is the production of distinctive forms that brings the phenomenon to the attention of an observer. Weight loss is a relatively simple index to measure, although it reflects gross losses and gain to a sample rather than just the affect of salt alone. Focusing on weight change as *the* index for salt weathering severely restricts both the types of experiments that can be run, and the type of questions that can be asked about salt weathering. The researcher will focus only on questions that can be answered by analysing weight change. Similarly, the whole experimental design will be dictated by the need to weigh and reweigh samples. This means that any design that does not involve repeated removal of the sample for weighing would not give results that are comparable to other experiments. Once a technique becomes established it is difficult for other techniques to develop to assess the same phenomenon. This is because the results from the new technique cannot be, except in certain circumstances, calibrated with the established index.

Moving to field-based experiments, such as erosion plots or weathering exposure trials, there is less control over the variables and their variations. Experiments, such as erosion plots, try to mix supposed control with the variability and uncertainty that is present in the reality. In erosion experiments, for example, variables that can be measured and whose spatial variability can be restricted are usually 'controlled'. Identification of variables and their properties is dependent on the ability of the researcher to measure them and to ensure that these properties remain constant or different from each other within experimental plots. The researcher has to be confident that the variable and properties measured do not vary spatially by a magnitude greater than the variability between plots. The researcher is relying on the spatial homogeneity of a variable or property to exist. Similarly, the researcher is relying upon any internal variability of a variable within a plot to have an impact upon the output measured that is less than the impact of the difference in variables between plots. The researcher is, therefore, assuming that they can determine the variability of reality at a scale relevant for their study and that they can also measure a meaningful output from the reality they have controlled. The researcher is relying on their measurement methods and theories identifying and being sufficiently accurate so as to constrain variation to a level that they deem sufficient for their analysis.

Monitoring of the environment suffers from difficulties as well. Although the researchers may not be trying to control reality, or even ensure a lack of spatial variability, their activity is still interfering with reality. The act of measurement is always an act of intervention. An example of this effect is the attempt to measure stream velocities. Inserting a probe into a flow will affect the flow itself. There is a feedback between the measurement instrument and the phenomenon that the researcher is trying to measure. Eliminating the feedback and the impact of intervention is difficult as the researcher cannot know what the phenomenon would have been if the intervention had not occurred. The only way to measure the phenomenon is with the instrument that causes the interference. The possibility of using a different instrument to measure the same phenomenon may be seen as a solution to this problem. This entails a major assumption however. It assumes that the phenomenon has an existence independent of the measurement instrument.

The above illustrates that the 'field' can be defined and constructed by the researcher. Significantly, the field of one researcher is not necessarily the same as the field of another researcher. Each researcher will have different theories to assess and define different boundaries to the world they want to assess. It could be argued, however, that the development of traditions amongst practising physical geographers means that there will be similarities in the bounding of reality. Clear identification of boundaries, as noted by Richards (1996) and Lane (2001), are vital for the development of intensive research. Whether such clear and consistent bounding occurs between researchers is unclear. World-making is a vital part of the research process, but it is often neglected (or rather ignored) by scientists, as it implies subjectivity in their world view. World-making does not negate objectivity but it does mean that researchers need to bear in mind the consistency of their world-making as the basis for their decisions.

Measurement systems

Measurement of phenomena requires the use of instruments to probe reality and impart quantitative information about its variability. Instruments are not selected in a vacuum. Although the use of a particular instrument in a particular type of study may become automatic, this does not mean that the use of the instrument is made without some regard to theory. Theory guides the researcher in terms of what to study in the physical environment and also what to measure.

Translating theory into something that can be assessed in the physical environment means that phenomena are made measurable. As with the field discussed above, it is often assumed that the phenomenon being measured is the same no matter how it is measured. Translations of theory into practice may differ, but these do not alter the nature of the phenomenon being measured. Without this assumption, measurement between different individuals made in different places or at different times could not be compared. The assumption does not, however, mean that this viewpoint is correct. Within quantum physics there is a view that the observer and the phenomenon cannot be separated. The observer and phenomenon make up a single system, a measurement system. In this context it is not possible to separate the measurement made from the measurement system within which it was made. The measurement and the phenomenon become combined in an unbreakable link in the measurement system. It is impossible to talk of a separate existence for the phenomenon and so also, therefore, to talk of an independent measurement of that phenomenon. This means that within the supposedly objective, hard science of physics, it is accepted that reality and how it is measured form an inseparable whole.

Rhoads and Thorn (1996b), based on the work of Suppe (1977), van Frassen (1987) and Giere (1988), put forward a view that the measurement of information in the environment cannot be divorced from the theory guiding the instruments used to measure the phenomenon. They summarised this view within the model-theoretical view (MTV) of scientific research (Figure 5.1). In this view a theory has associated with it a class of abstract structures, its models. These models are all derived from the theory and form a closely related family, related by the same postulates and laws. The models are intimately linked to their parent theory, but the models also need to be linked to reality in order to assess the theory. The linking is, as noted previously, achieved by the construction of hypotheses. These hypotheses are, however, related to the model, to the representation of the theory, rather than directly to the theory itself. Previously, however, it has been assumed that there is a link between hypotheses and reality. This link is more indirect. Hypotheses are linked to user-defined classes of phenomena that are thought to be present in the real world. Hypotheses are linked to data collected about the phenomena classes, not data collected about the real world as such. Data production is not independent of its own theoretical structures. Measurement of velocities within rivers by a laser Doppler or an acoustic Doppler monitor relies upon auxiliary theories in physics that link the change in wavelength of light and sound respectively with changes in velocity in a fluid. Likewise, these theoretical changes are only perceivable because they are 'sensed' by the instrument. This produces another layer of auxiliary

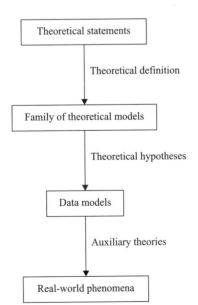

Figure 5.1 Model-theoretical view.

theories concerning the changes in voltage caused by changes in the environment external to the sensor. Similarly, there are a series of theories and protocols associated with appropriate methods of data collection and appropriate methods of data analysis. The layers are potentially infinite.

Combined, the auxiliary influences mean that there is a dynamic and ongoing process of construction involved in defining, collecting and analysing any data set. Rhoads and Thorn (1996b) call the aggregate of such a process a data model. The data model ensures that there is compatibility between the data harvested from reality and the requirements of the model to be assessed. As Rhoads and Thorn note, this means that:

> a theoretical model specifies the pattern of data a phenomenon or set of phenomenon should generate under a particular set of idealized conditions and the data model reveals whether this pattern of data is, in fact, present in the data collected from the real world.
>
> (Rhoads and Thorn, 1996b, p. 128)

This means that the theory explicitly guides the researcher and ensures that the data collected are appropriate for its assessment. Data are not raw material, ready for collection and divorced from theory. Data are defined and constructed by the theory via its theoretical models. The wonder is that the same data can be appropriate for the assessment of different models. This implies that the link to a specific model is not limited to a one-to-one relationship. Aspects of data are relevant to different models, potentially related to different theoretical frameworks. This may imply that data are real, that data do exist independently of theory. Alternatively, this property may reflect the limited manner in which models are perceived, and the limitations on what can be measured in the real world. There may only be a limited number of models that can be derived from a theory that are capable of assessment and of these most may require the same data for assessment. Similarly, what are appropriate data may be defined very narrowly by a specific field of enquiry. This means that a tradition or habit will prevail in what counts as correct data.

It should not be assumed, however, that data are static. The information collected by a single instrument may be capable of conversion into other data, other information, using different analytical procedures. Velocity measurements of a single point of flow may appear to produce no pattern when plotted against time. Their subsequent analysis using a running average may highlight patterns that were not visible before. Could you say that the same data were involved in each case? If you accept that data refers to the single measurement points then the answer is yes, but if you view data as the final information produced then the answer is no. Information is always capable of reinterpretation. Reinterpretation depends upon the development and application of new auxiliary theories and upon the impetus to extract such information based upon a new theory or new model that specifies different data for its assessment.

Rhoads and Thorn (1996b) highlight that the MTV view could be advantageous in physical geography. MTV puts the model, rather than the theory, at the focus of analysis. This is useful in seeming to mimic what actually happens in physical geography according to Rhoads and Thorn, but it is also problematic. The model is the thing being assessed and so it is also the thing being rejected or accepted. The status of the theory from which it is derived is always uncertain. Much as in Lakatos's research programmes, the central core can never be attacked, it can never be accepted as a firm or even a 'real' foundation for understanding. It is only the protective belt of models, in this case, that are exposed to debate. This does, however, mean that study tends to focus on the level of reality appropriate to the model. Assessment is being made of the model derived from a theory that is likely to use other theories based in physics and chemistry for its construction. These 'deeper' theories do not become the focus for rejection, only the models derived from a theory of which they are the building block.

MTV does not require that models be expressed mathematically. This means that cherished concepts in physical geography that are expressed qualitatively are acceptable as theories. This does not imply anything about what is acceptable as data within a data model however. Physical geographers may be happy to accept qualitative theories but may be more reluctant to accept qualitative information in their data models. Criteria of acceptance need not be the same for each part of the MTV process. Central to the MTV are the role of hypotheses, which many physical geographers have argued are a central aspect of practice (e.g. Chamberlin, 1890; Gilbert, 1896; Schumm, 1991; Baker 1996b). The application of MTV in biology, with its similar field-based phenomena, is a good illustration of the potential of the approach (Thompson 1983; Beatty, 1987).

Although the MTV has yet to be fully applied in physical geography, it does suggest a potential for developing an alternative but compatible framework for investigating reality in a field discipline relative to the 'hard' sciences. Physical geography has tended to 'borrow' heavily from physics and chemistry to justify its methods and practices. The difficulty of applying conditions of control and certainty of causality, however misplaced, to phenomena in the field has proved difficult, if not pointless. Physical geographers have tended, whether by design or by intuition, to employ a technique that they feel increases the certainty of their explanations: the principle of triangulation. Although the term is again borrowed, this time from surveying, it does act as a good metaphor for the development of practices for obtaining evidence in physical geography. Within surveying, triangulation refers to the technique for locating a single unknown point from different known points. Increasing the number of known points from which a single point is observed should fix that point's absolute location more and more accurately. There is increasing confidence in the belief that the point is accurately located the more surveying stations that are used to determine its location.

Triangulation within physical geography metaphorically refers to the use of different methods and different sources of information to assess the same model and, by implication, theory. This does not necessarily mean that different data models are being used. The same data model may permit different types of information to be collected, each valid as a means of assessing the theory. When the different methods or different sources imply that same result, then there is a convergence of evidence. When the different methods all diverge, they imply that there is no agreement about the status of the model, i.e. valid or invalid. Their divergence does cast doubt about the status of the model, and hence theory, thus increasing the level of uncertainty about its explanatory potential. An interesting situation may arise where there is both convergence and divergence. In this case, some methods may imply a similar result whilst others may imply a different, but consistent, result. In surveying, this would point to a systematic error in the locating of a point by certain surveying stations. A search would then be made for possible reasons why these stations produce such systematic variation. A similar conclusion might be drawn in physical geography. The presence of both convergence and divergence may point to a systematic variation in what each model, and the associated data model, explain or indicate about the phenomenon. Such differences may point to different theories and models illuminating different aspects of a phenomenon rather than being the explanation of a phenomenon.

Case Study

Triangulation of techniques – measurement of surface form on rocks

Analysis of the morphology of weathered surfaces has used a variety of techniques. Trudgill *et al.* (1989, 2001) used a micro-erosion meter to measure changes in the height of points on a surface relative to a reference plane. An indication of change for each surface was derived by averaging the height changes for 120 points over each time period. It was assumed that 120 points provided a satisfactory representation of the surface. Williams *et al.* (2000) used a laser scanner to characterise surfaces on wave-cut platforms. Their method collects tens of thousands of height points in each time period. Inkpen *et al.* (2000) used close-range photogrammetry to derive a photogrammetric model of rock surfaces. The number of point data on any particular set of images collected depended upon the requirements of the operator in analytical photogrammetry and the software in digital photogrammetry. Both papers assume that the surface they are trying to measure is real, but both also highlight that the techniques being used have problems, or rather limitations, in providing a 'true' measure of the surface. Any measurement technique carves reality up in a distinct and limited number of ways. The entity resulting from the carving of reality is not understandable outside of the context of its associated measurement technique. Linking entities between techniques implies an ability to agree on the presence of an identical carving of reality.

In both studies, the surface heights are only the final outcome in a complicated network of relations. Hidden behind these heights are a raft of other networks such as technical experts, commercial organisations as well as material networks such as electronics and other equipment. These networks are often hidden and become 'black boxes' of which the individuals measuring the surfaces know very little. With the laser scanner, for example, detailed knowledge of the electronics and how these influence the operation of the scanner may be required to interpret the behaviour of the laser. This is contained within the software used to analyse the returning laser light. With commercially available scanners, it is likely that few users would understand the equipment in such detail or even be aware of how different contextual conditions may influence results. An important part of these intertwined networks is the training required to produce the surface heights. Close-range photogrammetry, for example, requires the user to understand surveying techniques and photogrammetric techniques such as 'floating point' measurement of heights, and to establish appropriate control networks. The implementation of this training requires existing and often expensive capital equipment, such as the total station and analytical plotter. The recent and increasing use of digital photogrammetry to interpret stereo-images does not remove this knowledge, but again hides it within the software.

Each paper uses a different property of the surface to determine the nature of that surface. The laser scanner measures surface height by measuring the time it takes for light emitted by the laser to return from a surface. The return time is a function of the laser velocity as meditated through interaction with the target surface as well as its transmission through the instrument lens and detector. These data can then be related to the control frame or reference frame that each has used to then permit a calculation of surface heights. Close-range photogrammetry uses stereo-images of a surface taken with a calibrated camera as the data source. There are particular photographic require-ments that must be met to provide an adequate pair of images, for example at least a 60 per cent overlap between the images is required. The surface properties, as detected by the film and transmitted through the lens, are the basis of all further manipulation of the information. A long procedure of image interpretation is then required before a product of surface heights can be produced. With digital photogrammetry, there is often the additional step of scanning photography to go through, a step which can introduce another source of change into the information contained in the images. These differences in the routes used to determine the same product, surface height, can remain hidden once the data are presented and interpreted as a real set of surface heights.

Despite the similarity in the final output, surface heights, and the commonality of the subject area, both papers on measuring stone surfaces are associated with slightly different networks of relations that define their techniques. Both networks, however, provide for stable translations of instrument readings into representations of a 'real' surface. Similarly, both networks have sets of established common practices and are highly dependent on the crystallisation of network relations into durable and expensive equipment.

Both sets of height values are interpreted as if they refer to the same entity. Outputs from both measurement techniques are assumed to be different pictures of the same real surface. The differences between the two networks are reconciled through the idea that both papers and measurement systems refer to the same entity, surface height points. The unproblematic nature of these points hides the differences in training and errors. Differences in surface heights are usually interpreted as errors that displace the surface heights from their 'true' values. Assuming a common external reality removes the point measurement from the measurement system of which it is a part. Taking the point measurement out of this context means that it can be manipulated and interpreted independently of its origin and, significantly, its context. It might be argued that the various control networks and systematic testing of each technique means that the results produced are reliable and do refer to the same underlying reality. Likewise, comparison of the two systems by measuring a standard surface should illustrate their accuracies and errors. Such testing and control is, however, only ever made within the context of the measurement system itself. How can a standard surface be produced without a measurement system to define it as a standard surface? The techniques measuring the surface would still produce error terms specific to, and an integral part of, that unique measurement system. Although it would be assumed that both techniques provided the same data on the standard surface, it could equally be argued that each measurement system is measuring a property of the surface that they have defined as surface height. The surface and instrument properties that each measurement system is using to derive information could be different, but within our interpretative network they are lumped together as surface height. It is the assumption that real external surfaces possess a singular property called surface height that permits both the accommodation of the information from each technique into a common interpretative framework and the decontextualisation of the information about surface heights.

Practice in physical geography

Putting any theory and its associated models and data models into practice, actually assessing them, involves the development or rather construction of a location for that purpose. Location can refer to a laboratory, as much as to the field or experimental plot. The important point is that all locations are prepared in some way; they are not reality as it is, they represent reality as conditioned and constructed, often physically, by the researcher. This construction has been noted above, but the question becomes: how is this carried out in any study? Richards *et al.* (1997) and Richards (1996) provide some indication of the approaches available to physical geography. Richards *et al.* (1997) view two relationships as key in measurement, from the real to the actual and from the actual to the empirical. In other words, taking a realist perspective on measurement, the key relationships concern the interpretation or mediation between the levels of reality, the real, the actual, and the empirical. Richards *et al.* (1997) view physical geography as having vague and imprecise concepts such as equilibrium and, it could be claimed, more recently self-organisation, that are translated into entities capable of measurement, as noted in Chapter 2. They see a problem, however, in this imprecision. Imprecision means that there may be more than one translation thought to be acceptable and so different, distinct entities are classed as the same thing. Equifinality appears because of this imprecise translation. The second relationship, between entities and facts, they term transduction. Relating their definition to Pawson (1989), they view transduction as the conversion of an entity to a measurable quantity. The process involves conversion of one thing into another, or rather the representation of one thing by another. Different measurement systems could be used to identify and measure an entity, and all could concur that it exists and responds in a similar manner to each measurement system. In this case it could be argued that there is increasingly greater confidence in the 'real' existence of the entity. The idea is, as also noted by Hacking (1983), that if different measurement systems with different auxiliary theories identify and quantify the same thing and in the same manner, then there is likely to be something 'real' behind the measurements. Whilst this is a reasonable starting point, the researcher can never be certain that what the measurement system coincides or triangulates upon is the entity that they believe exists. It may be an entity, but its properties may produce effects beyond those measured, and its actions may only be measurable through its constant co-occurrence with other entities in the measurement contexts studied. These problems may become clearer if fieldwork is undertaken. This is another case of the importance of practice and theory informing each other.

Richards (1996), building on previous work, sets out the case for different types of research strategies within physical geography. He tends to build upon the intensive/ extensive division suggested by Sayer (1992) for use within the social sciences. In this framework, an extensive research strategy, which Richards broadly equates with more 'positivistic' science, identifies, defines and even quantifies relationships between entities by using a large number of measurements or samples. These are samples of reality. Samples are used in extensive research strategies to illustrate the product or the form of a variable. Samples are required in large numbers because they are used to define and quantify statistical relationships. More samples means greater confidence in the presence and nature of the statistical relationship. Extensive research strategies focus on gathering information on generalities, on patterns of forms. This requires a lot of information on these forms, and so requires a large number of samples. The samples studied possess the property in which the extensive research strategy is interested. Individually, samples may possess other properties that could vary between samples. Unless these other properties dramatically disturb the predicted pattern produced, they are invisible in the extensive research design. Extensive research is not interested in assessing the complexity or even complicated nature of the production of a particular form. Richards (1996) suggests that this reflects the usefulness of the extensive approach for positivistic and reductionist views of science. Patterns exist across different specific contexts. Data are collected concerning these patterns from different specific contexts and so they represent generalities that can be discovered in reality. The relationships that produce these patterns can be quantified and even modelled. Predictable outcomes from simple interactions of specific variables is the end result of extensive research. Why these patterns occur and why they occur in specific situations, and not others, is not really within the scope of an extensive research design.

Intensive research strategies are, according to Richards (1996), more in tune with a realist perspective on reality. Examples rather than samples are the focus of this strategy. Intensive studies deal with few, or even only one, detailed case study. The purpose of this case study is to identify and explore how mechanisms (the real) operate in a given set of circumstances (the actual) to produce measurable alterations (the empirical). Case studies permit analysis of phenomena across all three levels of realism's stratified reality. Central to the development of the case-study approach is a clear understanding of the theoretical aspects of a location. Abstraction of a theoretical basis for entities and interactions within a location provide the basis for any case study. From this framework, expectations of behaviour can be derived and applied to help measurement of the environment. Observations within the case study are the contingent outcome of the presence of general mechanisms and the coincidence of appropriate entities for the mediation of them. These entities exist only within a specific spatial and temporal context, the field location. Their mediation of mechanisms produces the range of measurable variables used in analysis. Mechanisms operate with differing intensities because of the uniqueness of the context, the uniqueness of the entity mix, with some mixes enhancing the action of a mechanism, whilst others inhibit it totally. It is only through an exploration of these different and unique contexts that the generality of mechanisms can be recognised and, importantly, the conditions of their operation even partially understood. This form of strategy is viewed as more appropriately suited to trying to understand detailed process-form relationships.

Case-study protocols have been commonly used within social sciences. Their role is much as identified by Richards (1996), the application of theoretical structures to specific situations or contexts. As in Richards's scenario, this application permits a greater understanding of the operation of a theoretical structure by identifying the parameters and limits of its actions. This strategy, however, relies upon the appropriate identification of a theoretical structure and its translation into a field location. Richards (1996) suggests that rather than field studies just being either extensive or intensive, it is more appropriate to view them as continually switching from one form to the other.

> However, these two styles of research, and the generalisations and explanations that they generate, are inextricably linked, and in any area of geomorphological enquiry there is a continual spiralling between them. This reflects a movement of the study from outside the case(s) in more extensive investigations, to inside a case in an intensive investigation (when new questions may be posed that require the embedding of additional extensive enquiries within the intensive case study).
>
> (Richards, 1996, p. 175)

The important point is that the two forms of strategies need to be combined to study physical reality. Constructing a theoretical framework for intensive study requires some initial starting point, often provided by information derived from extensive studies. Extensive studies can help to identify entities that may be of significance within a particular theory, that tend, for example, to co-vary in a predictable manner producing distinct patterns. By using this information it is possible to identify field locations where these variables are present and where their operation should be open to monitoring and further analysis. Vital to this latter task is the appropriate closure of the location. This means the identification of the system under study and its environment are significant preliminary stages of any investigation. Identification of boundaries is vital and also, importantly, why these boundaries are the focus of investigation. Likewise, the measurement of a range of entities and their values within the bounded location is important. Rather than just a single property, intensive study is interested in characterising the location. For this it is necessary to understand the detailed spatial and temporal variations in the location. The operation of a mechanism is likely to be reflected in, and impact upon, these variations in process and form. The purpose of the intensive study is to trace the mechanism in its mediation to the actual and empirical. This means detailed and wide-ranging measurement of entities and properties rather than just a simple single property measurement.

The analysis of the operation of mechanisms can only be made if the conditions of their operation are fully understood. Although this may be impossible to achieve practically, the narrowing of the potential uncertainties in the system can help to narrow the potential range of actions of mechanisms. It is important that the properties of any selected field location are outlined clearly and justified in terms of the mechanisms under investigation. In this manner it becomes clearer why specific results might be expected at these locations. Establishing the action of mechanisms at one location does not, however, imply that these mechanisms operate in a similar manner elsewhere. This is where further extensive studies, to identify locations with similar properties and with dissimilar properties, are necessary. Intensive study of the similar locations provides information on the general nature of the operation of mechanisms for a given set of conditions and how variation of these conditions affects the mechanisms. The dissimilar locations can help to assess whether mechanisms can operate even when specific conditions appear to be inhibiting. Additionally, undertaking an intensive study may require the use of extensive surveying techniques to analyse a particular aspect of the operation of mechanisms in a location. For example, in a detailed study of velocity structures within a stretch of river, intensive strategies involving monitoring of a range of parameters may still require the use of a sampling strategy to select appropriate points. The points sampled are then likely to require further statistical manipulation to produce a 'representative' numerical description of the flow. Although the statistical manipulation may be tightly constrained by established practices, the data collected and analysed still require both extensive and intensive methods of data collection.

Case Study

Linking process and form – intensive study of bedforms

Interpretation of process and form links in physical geography has often been viewed as a simple cause and effect relationship. Process is equated with cause and form with effect. The relationship is likely to be more complicated than this simple view. Lane and Richards (1997) assessed the linking of form and process within a specific river channel. Initial analysis of the data from an active braided reach of a river, the Arolla, in a glaciated catchment in the Swiss Alps suggested that there was a good correlation between the magnitude and direction of a change in discharge and the volume of material eroded or deposited. Large events produced the highest erosion. This observation would support a simple interpretation of magnitude and frequency relationships outlined by Wolman and Miller (1960). Closer analysis of the data suggested a more complicated relationship however. Sediment supply seemed to also influence the magnitude of erosion or deposition. Specifically, points later on in the rising limb of the hydrograph showed less erosion and even deposition in some cases, whilst the falling limb often showed deposition independent of the magnitude of flow. This suggested that both discharge and sediment supply controlled channel change. The situation was even more complicated. Sediment supply itself was determined by patterns of upstream erosion and deposition as well as local supply of sediment from bank erosion. Upstream sediment supply depends on the interaction of daily discharge with available sediment. The possible multiple combinations of discharge and sediment supply mean it is impossible to specify in advance what discharge will be the 'dominant' flow for producing channel alteration. The concept of a 'dominant' or formative event can only be meaningfully understood within the context of the whole catchment. Understanding a particular reach where measurements need to be taken cannot be divorced from the relationship of that reach to the rest of the catchment.

This view of magnitude and frequency relationships also meant that the response of a reach to an event could not be understood without reference to the history of that reach, the 'conditioning' effect of previous events. Lane and Richards (1997) suggest that history has a spatial manifestation in the morphology of the reach. Processes of sediment transport will be influenced by channel morphology and vice versa. As the channel morphology alters, so the sediment-transporting capacity of the channel will alter. Figure 5.2 illustrates how the patterns of erosion and deposition evolved in the reach. Initial deposition on both sides of the channel, with erosion confined to the scouring of a

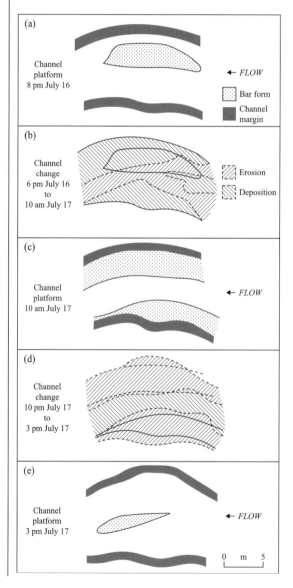

Figure 5.2 Alteration of erosional and depositional zones in a channel, from Lane and Richards (1997).

narrow channel, produces a morphology that will influence further zones of deposition and erosion. The lateral depositions acted as sediment sources for the next morning's flow. Consequent downstream deposition encouraged the development of a new medial bar. The sensitivity of different areas of the morphology to change varied as well. The fossilisation or otherwise of features depends upon their interaction with process events, that are themselves mediated by morphology. Lane and Richards suggested that the long-term development of the main medial bar was determined by the effects of processes operating at a shorter timescale and smaller spatial scale. These processes, however, altered as the morphology evolved. This means that process events cause change in the morphology of the reach, but in turn the process events are mediated by the initial morphology. The precise response of a reach to a process event will vary as the morphology evolves. History, as noted by Schumm (1991), is important for understanding physical systems.

Case Study

Probing reality – fluvial flow structure

The nature of river flow has recently been the subject of debate in the study of gravel-bedded rivers (e.g. Yalin, 1992; Biron *et al.*, 1996; Bradbrook *et al.*, 1998 and Lane *et al.*, 1999). Buffin-Belanger *et al.* (2000) set out to confirm the existence of large-scale flow structures and to develop a technique for improving the visualisation of these forms. From this technique, they suggest that the high-speed 'wedges' of flow display a complex organisation. This paper illustrates a number of points relevant to this chapter. The construction of the entity of study, the high-speed wedges, relies upon the use of specific detection methods and a specific theoretical framework. The new technique of analysis actively constructs the new entity of analysis, the velocity fluctuations in a space-time matrix. Finally, the outcome of their analysis provides a basis for further discussion of their constructed entities. The entities can be treated as if they were real and so debated amongst other researchers in the field. This does not mean to say that they are 'real', or understandable, when isolated from the measurement system, only that they can be treated as such for discussion.

Buffin-Belanger *et al.* (2000) begin by noting that Falco (1977) modelled a link between small-scale eddies and larger-scale flow structures. These forms could be identified by their velocities relative to the velocity of the flow as a whole; they were either higher or lower. Similar theoretical and empirical structures were identified by Brown and Thomas (1977) and Nakagawa and Nezu (1981). Importantly, these large-scale features have been identified by a combination of procedures involving multiple sensor arrays and visualisation techniques rather than single sensors. The flow structures are only identified when time periods are aggregated and compared to the average flow velocity of the river. The flow structure is an artefact of a comparative computational process. This highlights that the entities are represented only within computational analysis; without this computation and decisions about periods of time being above or below the average values, the entities would not exist within studies.

There has been a triangulation of different techniques upon what is perceived to be a real set of entities that seems to increase confidence in their reality. Kirkbride and Ferguson (1995) and Ferguson et al. (1996), for example, use Markov chain analysis on velocity data from three different heights in a natural gravel-bedded river to identify different velocity states. Roy et al. (1996) use a single-probe burst technique to identify flow structures near-bed and in the outer reaches of flow. The flow structures are assumed to be real and to possess certain properties that by definition they must have to be classed as flow structures. It is these properties that the measurement systems are designed to identify and quantify. The runs of higher and lower values than average have a pattern that reflects the theoretical structure of the entity, of the flow structures. The convergence of different measurement techniques on the presence of these runs of values is viewed as evidence for the reality of these entities. The measurement systems used, however, are all designed to identify the units that make up the entities, the velocity measurements of a specific time slice.

Buffin-Belanger et al. (2000) measure their velocities in a straight section of the Eaton North river, Quebec, Canada at relatively low flow. The river had a poorly sorted gravel bed with a D_{50} of 33 mm and a sorting coefficient of 2.9. Mean flow depth was 0.35–0.40 m and mean flow velocity was 0.36 ms^{-1}. A beam was stretched across the river and a movable support arm attached to this. Three Marsh-McBirney bi-directional electromagnetic current meters (ECMs) were fixed 8 cm apart on a wading rod attached to the support arm. Each probe had a sensing volume of 3.25–3.9 cm^3 and a low-pass RC (resistor-capacitor) filter with a time constant of 0.05 s. These sensors were arranged to permit three simultaneous measurements of streamwise and vertical velocity components. This simultaneous measurement of these components was essential for the latter processing of the information to derive representations of the flow structures. The data model even at this stage influences the nature of the measurements taken. Twenty-three combinations of three simultaneous velocity measurements along five vertical profiles were taken along a 1.3 m transect parallel to the average flow streamline. This means that the latter condition had to be identified in the field and the equipment moved to be parallel to it. This restricted the experiment to a relatively straight stretch of river as already noted. This could restrict the phenomenon being identified to only these experimental conditions, rather than to the identification of a universal feature of all stretches.

In each vertical profile, initial velocity measurements were taken at 3.5, 11.5 and 19.5 cm with additional profiles being taken at 2 cm intervals up the rod until a height of either 25.5 or 27.5 cm was reached (the water surface). This experimental design produced 12–13 elevations with a total of 69 points of velocity measurements. At all points the streamwise (U) and vertical (V) velocity components were sampled for 60 s at a frequency of 20 Hz. In addition to these data, additional information was obtained from a triplet of measurements over a 20-minute period at a frequency of 20 Hz. This permitted an assessment of the temporal variability of the flow structures identified.

The above detailed description of the location and method provides information on the requirements for identifying the phenomena. The entities trying to be identified are not obvious from visual examination of the river. Instead, the entities are believed to exist from theoretical considerations, but their presence can only be confirmed by their signal in a specific and well-controlled set of information. This means that a detailed and well-confined experimental design is required to generate the quantity of information that can be statistically processed to derive appropriate representations of the entities. Do the phenomena exist outside of these carefully constructed experiments? For discussion of the entities, it has to be assumed that they do, but argument about the nature of these entities requires the same painstaking control of the environment to glimpse the entities.

Analysis of the data takes a similar approach of triangulation. Many statistical interpretations of the data are used rather than a single one to obtain different representations of the entities. Space-time correlation analysis (STCA), based on the calculation of correlation values between samples at different spatial locations and at different time lags, is used. From STCA, the maximum correlation values (r_{max}) and the time lag of this (L_{max}) are key derived properties of flow. A multisignal detection technique is also used based on computing the joint time series by averaging instantaneous velocity fluctuations for the three signals over a time period (1 s in this study). Flow structures are detected using the U-level technique (Bogard and Tiederman, 1986). This is based on the magnitude of the deviation of an instantaneous velocity value (standardised) from the average value of the time series. High or low flow velocities are identified relative to some threshold value.

The key innovation is, however, the use of a novel method for visualising flow structures. Velocity fluctuations (deviations of instantaneous velocity from a time-series average) are represented for each height as either a positive (black line) or negative (white line) deviation (Figure 5.3). The resulting matrix provides a visual representation of the persistence or otherwise of flow structures (Figure 5.4). Buffin-Belanger *et al.* suggest that the matrix is like an unrolled film beginning at the right edge and ending at the left. Visually it seems similar to a bar code and could be seen as representing the 'signature' of a river's flow over a given time period. With this method, large areas of black or white represent the persistence of flow structures, and the timing of these structures or wedges between the sensors can be determined thus identifying the characteristics of the shape of these flow structures. Refinements can be made to this matrix by adding thresholds for low and high flow enabling 'noise', small fluctuations, to be downplayed relative to the large fluctuations at the core of the high and low velocity wedges.

The whole paper is concerned with being able to identify and visualise the flow structures. The structures as such cannot be sensed directly but what can be sensed are the expected properties of these entities. Changes in velocity are a representation of these entities, although they are only a single, specific property associated with them.

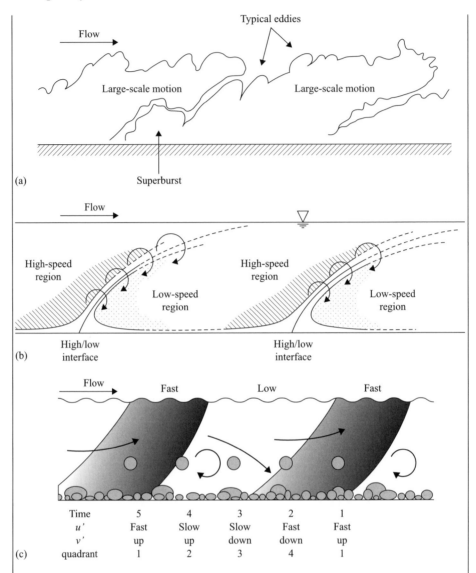

Figure 5.3 Flow structures, from Buffin-Belanger *et al.* (2000).

To this single measurable property a range of statistical techniques have been applied that seem to converge on both the identification of the entities and on their nature. Whether the convergence is because the entities are real, or because the measurement systems are searching for the same type of change, is irrelevant. The information is treated as if the entities are real and have a representation in the data. Flow structures are constructed by both sensors and statistics, but their reality is no less 'real' because of this. Their reality lies in their use as objects for discussion and analysis, objects that can be identified by others with measurement systems and their nature contested and tested by others.

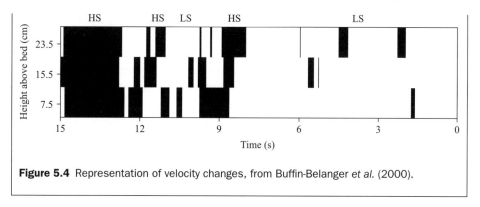

Figure 5.4 Representation of velocity changes, from Buffin-Belanger *et al.* (2000).

A key approach to investigating reality is provided by the method of multiple working hypotheses (MWH) (see examples below). The method of MWH has been championed by many authors (e.g. Haines-Young and Petch, 1986; Baker, 1999) and is related to the critical rationalist view of science, which holds falsification as the most important aspect of the scientific process. The approach takes a number of hypotheses and tests each in turn by finding a 'critical' test to apply so that that hypothesis can be excluded as a potential explanation. Identifying the failure of a hypothesis, however, is not a simple task and requires a consideration of the context of the researcher including the models and theories used (such as MTV, described previously) as well as the measurement systems available to them, which are all part of hypothesis creation.

Case Study

Multiple working hypotheses

Gilbert and crater formation

An early paper to explore this method was by Gilbert (1896) in relation to what he described as a topographic problem. Gilbert called a hypothesis a 'scientific guess' (1896, p. 3) and viewed the initial guess as the first step an investigator made to connect a group of facts and a cause. Rather than starting with a blank canvas, Gilbert saw the hypothesis as linking existing cause and effect. The role of the hypothesis was to provide a means of extracting information that would assess the validity of the link between the two. Gilbert reasoned the link as follows:

> In other words, he (sic) frames a hypothesis or invents a tentative theory. Then he proceeds to test the hypothesis, and in planning a test he reasons in this way: If the phenomenon was really produced in the hypothetic manner, then it should possess, in addition to features already observed, certain other specific features, and the discovery of these will serve to verify the hypothesis.
>
> (Gilbert, 1896, p. 3)

The hypothesis suggests new features, new observations that should exist if it is true. The investigator searches for these new features and if they are found the theory is supported, if not it is rejected. Gilbert states that a theory is accepted as satisfactory if the new features found to support it are numerous and varied. Gilbert does not seem prepared to accept a single new feature as sufficient grounds for accepting a theory. Gilbert states that a series of trials will result in inadequate explanations being set aside until one is found that satisfies all the tests carried out. Unlike Popper, Gilbert views this method as leading to verification of a satisfactory theory rather than non-falsification.

Gilbert then applies his technique to the problem of a crater at Coon Butte in northern central Arizona. Gilbert proposed four hypotheses that could explain the crater: explosive formation, meteorite impact, volcanic intrusion and volcanic explosion. Gilbert assesses each hypothesis in turn, using existing evidence in the case of the explosive formation and volcanic intrusion, and then outlining the critical testing he undertook to assess the remaining hypotheses. In each case, Gilbert outlines a logical, deductive sequence of events that could have resulted in the formation of the crater under each hypothesised cause. Associated with each sequence was a set of features that should be present if the sequence occurred. Data collection and analysis is then focused on identifying the presence or absence of these features for each hypothesis. This means that data collection and analysis is always done with a clear purpose, never for its own sake.

The explosive formation was initially supported by the circular shape of the crater and by the presence of iron, which was thought to have been ejected by the explosion. Analysis of the iron suggested it was of extraterrestrial origin, so eliminating the explosive hypothesis. The volcanic intrusion hypothesis relied upon the presence of a laccolite under the crater. This intrusive form would have forced the limestone and sandstone strata to deform as a dome. The crater would represent the remains of this updoming. The lack of any volcanic rocks within the crater eliminated this hypothesis.

The last two hypotheses were assessed by Gilbert in the field himself. He devised two critical tests, highlighting his concern that no single line of evidence is sufficient to verify a hypothesis. First, a detailed topographic survey should indicate whether the volume of material in the rim matched the volume of material excavated from the crater. Gilbert believed, from experimental work with projectiles, that a meteorite of a volume between 6.0 and 7.5 million cubic yards was necessary to produce the crater. Absence of any difference in volumes between the rim and crater would indicate that no such additional mass was present. Second, Gilbert carried out a detailed magnetic survey reasoning that an iron meteorite should leave a high iron content in the crater that would produce a deflection of a magnetic needle. The topographic survey found the same volume of material in the rim as excavated from the crater, 80 million cubic yards. Likewise, the magnetic survey found no deflection within the crater relative to the plain upon which it was located. Both these critical tests meant that the meteorite hypothesis should be rejected. Although Gilbert does reject the hypothesis, he appears reluctant to do so. In the text he states this was his favoured hypothesis before fieldwork began. Even when the data confronted him, Gilbert resorted to experimental work to assess whether the meteorite could have been smaller or at a depth where its magnetic influence would not be detectable at the surface. Although he does not accept the meteorite hypothesis, his comments leave it as an uncertain rejection rather than condemned outright.

The remaining hypothesis is not contradicted by the field data, but neither does it seem to be critically verified. Gilbert argues by analogy with other craters that are thought to have been formed by volcanically-induced steam explosions. It is important to recognise that Gilbert's analysis does not just rely on the field data. Gilbert switches between argument by analogy with other known forms to experimental work on processes such as projectile impacts. He uses the information from each to try to justify his deductive arguments about what features should be expected for each hypothesis. If the links between experimental work or analogy and his argument are later shown to be erroneous, then his chain of reasoning breaks down. The results and rejections rely on his reasoning being valid; if it is not, then the status of each hypothesis could alter.

Loss of coastal wetlands

More recently, a multiple working hypothesis framework was adopted by Turner (1997) in an attempt to identify the cause(s) of the dramatic loss of the coastal wetlands in the northern Gulf of Mexico. Between 1955 and 1978, almost 1% of the coastal wetlands of this region were lost per year; this equates to an area of wetland equivalent in size to Rhode Island being lost in a little over 20 years. Approximately 12% of this wetland loss was due to direct human activity (direct modification resulting from installing pipelines and canals for example). However, almost 90% of this wetland loss is believed to be due to other, indirect, causes. There was little consensus of opinion over the indirect cause(s) of this decline in wetland area. For example, the Louisiana State Government believed that a decrease in the suspended sediment load from the Mississippi River drainage system, following dam and reservoir construction on major tributaries for example, was the cause of wetland loss, and so restoration efforts should focus on sediment management. However, other causal mechanisms exist that could potentially explain this loss, such as an increase in wetland salinity arising from the presence of dredged channels. This would facilitate greater saltwater intrusion into the coastal zone. An inability to confidently identify the cause of this loss of wetland necessarily hinders attempts to conserve and restore these valuable ecosystems. Turner maintains that the lack of clarity in identifying the relative importance of the different mechanisms proposed to account for dramatic wetland loss arose because clearly stated hypotheses of potential causes were never formulated or rigorously tested. Turner (1997, p. 4) examined four competing hypotheses, or causal mechanisms, that have been proposed for the loss of wetlands (specific examples of each of these causal mechanisms leading to wetland loss exist for different portions of the coastline):

Hypothesis I: the indirect consequences of dredged canals and the resulting spoil banks have led to the majority of land loss since the 1930s.
Hypothesis II: A decline in suspended sediments of the Mississippi River during the 1950s led to the majority of the land loss.

Hypothesis III: Mississippi River navigation and flood protection levees, mostly built during the 1930s, led to the majority of the land loss.

Hypothesis IV: Saltwater intrusion from offshore to inland has caused the majority of land loss since the 1930s.

Sediment compaction and sea-level rise were believed to have been negligible during the interval under study and so were not considered. Reference should be made to Turner (1997, p. 4) for a detailed rationale of each of the hypotheses proposed.

Turner adopted a statistical model to quantify annual changes in wetland areas based on five land inventories of eight estuarine systems in Louisiana for the period 1930 to 1990. Model estimates of wetland loss due to direct impacts were verified by comparison with records held of dredging activity by the state's Department of Natural Resources. Knowledge of direct wetland loss allowed ratios of annual indirect:direct land loss to be calculated during the study interval. Using this approach, Turner was able to reject hypotheses II, III and IV. Hypothesis I could not be rejected. This hypothesis proposes that direct land loss, primarily through dredging, results in hydrological change and leads to additional land loss indirectly. A linear relationship was found to exist between the indirect and direct land loss. Linear regression of this relationship resulted in a zero intercept suggesting that other causal mechanisms were insignificant. Therefore, Turner argues that the indirect consequences of dredging could account for almost all of the wetland loss experienced on this coastline during the last 60 years. Furthermore, a link between wetland loss and dredging was supported by existing experimental field and laboratory studies. Turner concludes that, based on this study, many management models incorrectly emphasise terrigenous suspended sediment supply as primarily controlling wetland gain, loss and maintenance.

As Chamberlin (1890) states, the purpose of adopting a multiple working hypothesis approach is to investigate every rational explanation for a phenomenon. This approach is designed to enforce objectivity – research is not designed in a way to support a pet hypothesis or a prevailing view. If this were the case, Turner may have adopted a different approach and investigated exclusively changes in suspended sediment flux, rather than investigating a range of other possible causal factors. A further benefit of investigating several hypotheses in tandem is that more than one causal factor may be identified, and potential interrelationships between causal mechanisms may be uncovered. On a more practical note, however, a multiple working hypothesis approach may be beyond the capability of the research programme, perhaps due to time or cost constraints. This will particularly be the case where there are many competing hypotheses, and/or adequate research into each is time-consuming or expensive.

Summary

Dialogue with reality requires analysis in the field. The 'field' can be defined in different ways and reflects, to some extent, the different ways in which different physical geographers construct the reality they are studying. Dialogue requires intervention. Intervention means a researcher has to frame reality, has to decide what to include and what to exclude from their consideration. Examination of reality requires the use of measurement systems to provide information about the nature of phenomena and their variations. All measurement will be theory-ladened, but this does not automatically mean that all measurement is subjective. The model-theoretical view illustrates the intimate link between theory, data models and hypotheses. Theory guides the researcher as to what data to collect and in ensuring the data are compatible with the identification and quantification of the type of phenomena the theory expects to be present.

Triangulation through the use of different instruments measuring the same phenomenon can increase the confidence researchers have in the reality of that phenomenon. In such cases, the basis of each measurement system needs to be considered in detail as they may be based on the same principles rather than reflect different data models. Increasingly, the complexity of the physical environment is being recognised as extensive studies of entities and relationships yield little in the way of explanation. Extensive studies focus upon statistical relationships and the identification of generalities and patterns through the use of a large number of contextless samples. Intensive studies put entities and relations back into their context. These types of studies are concerned with why patterns occur in specific situations and not in others. Intensive studies identify and explore how mechanisms operate in a particular set of circumstances to produce empirically measurable change.

Chapter 6

The field

Physical geography has always considered itself a 'field science'. What this term actually means though has not really been explored in any depth, as noted by Driver (2000, 2001) in relation to geographical fieldwork in general. This chapter explores the term and highlights the importance of this self-definition for the academic and professional kudos of the subject. Specifically, this chapter tries to define what a 'field science' is, partly by looking at what the term attempts to exclude when physical geographers wander off to do fieldwork. This chapter then uses 'the field' as the starting point for the simplification of reality within the laboratory and as the arena within which monitoring occurs.

What is 'field science'?

The term 'field science' is often used to describe the difference between physical geography and the 'hard sciences' of physics, chemistry and, to a lesser extent, biology. The term has also been used, however, in relation to anthropology (Gupta and Ferguson, 1997; Robben and Sluka, 2012) and for the supposed 'hard' science of biology (Kohler, 2002, 2012). This suggests that whatever it is that physical geographers feel makes their subject unique, is also shared by other academic subjects. Searching for a commonality in the approaches, philosophies and methodologies of these diverse academic subjects does not seem to produce a set of specific definitions or philosophies that unite their research. The uniting factor seems to be a view that the focus of their research should be reality, the real world, as presented to them by 'the field'.

'The field' is viewed as having an existence independent and separate from the researcher. The field is out there and it is waiting for us to discover it. Phenomena exist in a raw state in the field, interacting and producing expressions of their existence, and it is up to our ingenuity as field scientists to uncover or, if you prefer, to discover. Discovery is this act of interrogation. We go into the field, we observe, we probe, and then we ask questions and reality answers us. The metaphor of researchers undertaking some sort of dialogue with reality, as mentioned in Chapter 5, is central to this view of the field. The field is the arena in which researchers and reality encounter each other.

Discovery can be a case of prising information from a reluctant reality. In this metaphor, the field is something to be tackled and tamed, something to be conquered.

Exploration and discovery are very aggressive, penetrative activities, leading some authors to take a very gendered view of the whole fieldwork experience. Rose (1992, 1993, 1997), for example, views the act of physical fieldwork as a very masculine undertaking. Fieldwork involves confronting nature, 'toughing it out' in the field to prise understanding from nature and so dominate it. The history of geographical exploration as a series of derring-dos does little to nullify this view. This view of field-work is not dissimilar to the view and approach adopted by Francis Bacon, the father of experimental philosophy, in which nature is contained and controlled through experimen-tation in order to extract its secrets (Merchant, in press). This view has attracted criticism, particularly from feminist philosophers and historians, because of the apparent emphasis in Bacon's approach on the domination and control of a nature gendered as female (e.g. Merchant, 1980), and the sexual connotations that this implies (Harding, 1986). Discovery can, however, also be viewed as a very creative act. Although discovery is hardly ever passive, the researcher could be seen as teasing and coaxing the expression of phenomena from reality rather than shaking these expressions from the field. Fieldwork requires the movement of the researcher into and through the field, the study of reality through movement within it (Outram, 1996, 1999). Movement causes change and, as noted by Driver (2001), this results in the field always being in a state of development, in the process of construction and reconstruction by the fieldworker. In addition, the phenomena being researched is always in a state of development, being altered by the processes and phenomena the researcher wishes to study as well as by those of which the researcher may have no knowledge or considers irrelevant. Entering the field changes it. By how much, in what way, and the degree to which this change can influence subsequent field observations and measurements, is part of the research practice of the discipline (Powell, 2002).

The 'hard' sciences are caricatured by physical geographers as being concerned with the physical and chemical basis of phenomena. Within critical realism this might be equated with the 'how' questions of mechanisms and processes. These sciences deal in the development of underlying theories that explain a whole family of phenomena. Importantly, the concepts from these sciences are refined and abstracted so much that they can be expressed clearly as mathematical relationships. Similarly, these phenomena are capable of detailed and repeated study and testing as they can be created and recreated as exact facsimiles in the laboratory. The 'hard' sciences abstract and understand phenomena from their context of production and observation, which is from 'the field'. So however much the 'hard' sciences may wish to be seen as isolating 'natural' phenomena, even these are intimately linked to the 'field'. Thus abstracted, the phenomena of investigation in the 'hard' sciences have no history and shun any hint of uniqueness in their explanation. The 'hard' sciences instead are concerned with the development of universals – universal laws applicable to universal entities that require no further investi-gation or explanation once these laws are established.

Although very much an unfair caricature of the 'hard' sciences, physical geography has increasingly viewed itself as practicing something different. The 'universals' are still at the heart of understanding processes in physical geography, but the entities of study are anything but universal. Physical geography grapples constantly with the issues of 'uniqueness' and 'history' of its entities of study and their context. These two properties provide the basis for the distinctiveness of a 'field' science as used in geography. The entities of research evolve and change as processes affect them and, in turn, alter their own context (Inkpen and Petley, 2001; Inkpen, 2007). The nature of this change is not necessarily predictable. The history of an entity and its context are an important part of the explanation of that entity. Although the entity of research may be sculpted by processes based on the 'hard' sciences, the outcome is a result of the unique combination of processes and forms in the setting of the field. Replication of this uniqueness in its entirety is impossible. It is not possible to bring the whole context of entities being researched back to the laboratory or even to the virtual confines of the computer hard drive. Reality cannot be replicated and the histories of an entity 'rerun' to assess whether the outcome was predictable from first principles or the universals of the 'hard' sciences.

This view of field sciences as dealing with reality in all its complexity is a romantic and a heroic one; whether this is mirrored by the reality of research is another matter. Moreover, the view that, unlike the physicists and chemists sheltered in their laboratories, the field scientists have to deal with reality in the raw, and extract understanding from the unique and complex entities that are the focus of research, is also mistaken. Field scientists never go out into the field and research everything about an entity or about the context of that entity. Research is a process of selectivity. Researchers, as outlined in the previous chapters, always go into the field with some theory to assess. Theory guides practice, and dictates which variables are selected for measurement by the researcher, and the type of behaviour expected to be witnessed. Their training, research traditions and questions determine what phenomena field scientists deem worthy of research and so limit their view of the 'field' to the expressions of these phenomena they believe they can witness. This means that the 'field' as a totality, as a complex holistic thing, is never the focus of research. It is a subset of this totality, selected, or rather bounded, by the researcher that forms the focus of research.

The observant reader will notice that we have tended to equate 'the field' with a notion of going outside to do fieldwork, of having to physically remove oneself from the comfy confines of the office and trek into the hot and sweaty (or in our environment of the UK, the wet and windy) 'outdoors'. This is too restrictive a view of 'the field' and limits the importance of places where geography is done. Wandering off into the 'great outdoors' to undertake, depending on your viewpoint, a battle or dialogue with nature neglects the other spaces and places that make this wandering possible. Undertaking the actual field-work can be viewed as a part of the process of field-based research. There are a number of stages involved in this process: planning and preparation, exploring the field, and post-field analysis. Each of these stages can be subdivided, if you so wished, into smaller, or specific tasks that help to define a particular piece of fieldwork. Planning and preparation can take place within a number of settings; the important aspect of this stage in the process is that it concerns the setting and framing of the research questions to be asked in the field and, from these discussions, the nature of the answers is determined. An office, an archive or a group meeting could all play a role in the development of key questions to be asked in the field and the determination of how these questions will be tackled. These discussions, personal and internal or group-based and external, finalise the loca-tions of research, the methods and equipment to be employed as well as the expectations of results. All these issues will be carried out within, and guided by the context of, the research topic. Whether you view these as paradigms or research programmes is not a concern, the important thing is that they will affect the nature of the questions asked and the phenomena, and hence field locations, to be investigated.

The logistics involved in organising and co-ordinating research questions, equipment, travel and, increasingly in physical geography, research teams, should not be under-estimated. This is an essential part of fieldwork as much as the intellectual exercise in developing research questions. These practical considerations will influence the nature of the research questions as well as the location of 'the field' to be researched. Equipment will need to be transported safely as well as operated effectively in the 'field' selected. The important point is that the 'field' is never selected without considering the logistics of actually asking the research questions. Once again this can constrain the type of locations selected for fieldwork.

Exploring the field is the next stage of the process. Once in the field, researchers then have to identify and monitor the entities of interest. Bounding and isolating phenomena is a skill that once again refers back to the traditions and research context of the researchers. Exploration isn't an expansive practice; it is exploring to restrict the scope of the field. In a familiar research site it is likely that researchers will develop tunnel vision in terms of what entities they can see. These will be predefined by their planning and preparation and will be the focus of their activities in the field. In an unfamiliar location, these same patterns of entity recognition and bounding will be overlain on the field site, turning the unfamiliar into the familiar. Exceptionally, researchers will try to examine a location with 'fresh eyes'. New theories, new instruments, even multidisciplinary teams, can aid in this fresh vision, as these all bring a new framework for seeing and interpreting a location.

Post-field analysis is a fairly bland term for the whole range of post-processing and interpretation that follows the return from the field. The places of post-field analysis can vary greatly as can the temporal distance between the fieldwork and the analysis. In some cases, researchers may simply retreat to a tent or to a field centre with a nice mug of coffee to process and interpret the day's field data. In other cases, extracting information may involve a long, often tedious, set of procedures such as in the analysis of pollen records in extracted cores. Whatever the degree of separation between the field and post-field analysis, the two activities are linked by the research questions they are meant to inform. This highlights the interconnected nature of the stages. Within post-field analysis, the nature of the field is open to redefinition and reinterpretation. This stage is also one in which researchers who have not been to a field location can become involved. In environmental reconstruction, for example, researchers will often send material to other researchers to analyse specific environmental proxies. These researchers need not have been in the field location collecting the sample on which they now work. The 'field' and all the decisions and issues bound up with it are still relevant to these researchers however. These issues have been vital in determining, and it might even be said in constructing, the material they are now studying. The issues, however, now become invisible, hidden through the acceptance that the material has been collected using accepted methods and so conforms to the norms of practice of that discipline for dealing with the field.

The philosophies of fieldwork

There is, however, a difference between the 'field' science and the 'hard' science study of entities and phenomena, in the philosophies employed to understand these, and in the nature of causality identified. Although the field scientist still selects what to study, the basis of their understanding of these phenomena is not limited to the processes uncovered by physics and chemistry, the basis of a purely reductionist explanation. Field science could be viewed as taking an approach to explanation that is more in line with a critical realist philosophy. Explanation is focused at the level or scale of interest rather than trying to force explanation down to the level of the basic principles of physics and chemistry. Trekking into the field implies that the researchers view the entities and phenomena of interest as only being identifiable and able to be studied within the particular context of the 'field'.

Within critical realism the stratified and differentiated nature of reality means that entities can be identified at the level of interest, placed within the context of the level above them and explained by reference to processes and entities identified at the level below this level of interest. Reference to the 'lowest' level of the 'hard' sciences for explanation is not warranted within critical realism (although it could be argued that even this level is not the lowest as quantum mechanics forms the explanatory basis of the supposedly fundamental sciences of physics and chemistry). Explanation and causation is reduced to the appropriate level – the level below the level that is the focus of research. This means that fieldwork is an essential way to identify and analyse relevant entities and relevant processes for explanation and understanding, as it focuses research on the appropriate stratified level of reality for the phenomena under study.

Fieldwork is also essential for identifying causality. In Chapter 4 the views of Cartwright on causation were outlined. Two key issues were the unstable enablers and external validity (Cartwright, 2007; Cartwright and Efstathiou, 2011). Unstable enablers result from changes in enabling factors that support causal laws whilst external validity results from the problem of trying to generalise from a particular setting or population to another of interest. Cartwright and Efstathiou (2011) redescribe these issues as being issues of time and space. Unstable enablers are concerned with changes in known and unknown structural factors through time. External validity becomes a concern when there is a change in the case being considered, when the results and insights from one case study in a specific place are used to infer similarities to another case study in another specific place.

> The problem of 'external validity' speaks of changes as arising when we move in space, from places 'interior' to those 'exterior' to our case study. So we might say that the problem becomes relevant when we think of causal knowledge as shifted across space rather than when things change with time.
>
> (Cartwright and Efstathiou, 2011, p. 232)

This viewpoint suggests that fieldwork is a vital component in ensuring that the nature of both unstable enablers and external validity is identified and made visible in the development of explanations for phenomena.

Both Cartwright and critical realism are concerned with identifying causes. Both, however, are not searching for universal causes. Cartwright (1989) views causes as relative to the nomological machine, the underlying structure, and to the particular arrangement of confounding factors that they occur within. Keeping these two elements the same produces causal regularity. The need to ossify these elements, however, makes the identification and continued operation of causal regularity very fragile. In the field, such ossification is unlikely to happen unless humans manage it.

Instead of hunting for causes that operate regularly, critical realism searches for capacities and dispositions. Capacities are the powers that entities possess to contribute to the results we observe. It is the capacity of an entity to do something that is what is observed in the field. This capacity is not, however, realised to the same extent in every set of conditions. Research can establish that capacities exist, but it is only through the realisation of a capacity that its presence can be identified and the power of a capacity to cause change or maintain a situation analysed. The realisation of a capacity and the identification of this realisation require fieldwork. It is only through the analysis of the capacities of an entity in interactions with other entities that the range of possible outcomes and relative contributions of an entity in a specific configuration can be understood. It is the very variability of the field which seems to preclude it from scientific study that is its greatest strength for understanding how scientific principles can explain reality.

Turner (2005) and Cleland (2002) provide some interesting insights into the nature of causation and explanation in field sciences, although they discuss the issues in terms of 'historical' and 'experimental' sciences. Both authors focus on the issue of 'overdetermination' as identified by Lewis (1979) as a key difference in both approach and purpose of 'historical' and 'experimental' sciences. Field science covers both aspects (see section below) but in trying to explain phenomena in the field, the issue of over- and underdetermination is a key issue that is often unrecognised at least explicitly. Overdetermination refers to the issue that any fact about the world can, according to Lewis (1979), be determined by a minimum set of conditions along with a given law of nature relevant to the fact under consideration. The same fact, or event, can be overdetermined if it has more than one determinant at any given time. Lewis then states that overdetermination is asymmetrical, with earlier events being overdetermined by later events. Cleland (2002) suggests that historical scientists exploit this overdetermination as they expect past events to have a large number of determinants in the present, where determinants are taken to be sets of conditions or 'traces' of the past event that, when combined with relevant laws, are together sufficient to explain the earlier event. The example of the extinction of the dinosaurs by an asteroid 65 million years ago is used as a classic example of overdetermination by Cleland. The Chicxulub crater off the coast of central America (Hilderband et al., 1991), the iridium layer at the Cretaceous-Tertiary boundary (Smit and Hertogen, 1980; Nichols et al., 1986) as well as the presence of shocked quartz at the same boundary (Bohor et al., 1987; Bice et al., 1992; Alvarez et al., 1995) are all taken as indicators of the occurrence of an extinction-causing asteroid impact (Alvarez et al., 1980). The past event has an abundance of traces in the present that point to the occurrence of that past event and no other event that can match the traces – Cleland's 'smoking guns'.

The future and present, however, are not overdetermined, in fact they are underdetermined. Any event observed in the present has a number of possible causes or explanations that the researcher has to narrow down to determine which are relevant. Cleland suggests that this is where more classic 'experimental' science is used to prise understanding from nature. In this approach the researcher tries to form a hypothesis based on observations, which is then tested by controlling or manipulating nature. This control or manipulation is through the controlling of variables. Controlling variables reduces the number of possible causes that could determine the observations. Control, however, also limits the nature of the events observed and the researcher needs to run a whole host of controlled experiments to decide which manufactured events really match the observed patterns, and which of the manufactured events are merely artefacts of the experimental procedures.

Turner argues that Lewis is thinking about overdetermination as a metaphysical idea relating to the relationship between earlier and later events, whilst Cleland is thinking of overdetermination in terms of relations between hypothesis and evidence. Turner suggests that overdetermination may be an appropriate way of viewing the past in general; it does not rule out local underdetermination of events. He suggests that through time the multiple traces (e.g. physical, chemical and biological proxies) of a past event do not remain stable in the environment. Information-destroying processes (Sober, 1988) remove or distort traces of past events and as such there may be few traces left in the present to determine with confidence the exact nature or occurrence of a past event. It should be noted that processes that remove traces are not necessarily the same as those that distort traces. Removal implies that the trace is missing when it would be expected. Distortion can be of two types. First, that the trace has been altered and so it can be interpreted in a number of different ways in relation to a past event or events. Second, that the trace may have remained unaltered, but how to interpret the trace is unclear. As highlighted in our discussion on abductive reasoning (Chapter 4), without other traces in the environment an individual trace can have multiple meanings in relation to a past event or events.

Local underdetermination means that a researcher will have to weigh the evidence for competing hypotheses in trying to understand the past but without the ability to choose between these hypotheses given the available evidence. (Turner cites this as incompatible hypotheses being empirically equivalent in the weak sense.) In this case it may be possible to refer to underlying theories about the past that the hypotheses are derived from to assess which is more likely. If, however, these underlying theories suggest that the hypotheses are equally as likely then the hypotheses are also strongly empirically equivalent and the task of choice is near impossible. Turner suggests that this issue of local underdetermination is widespread in historical sciences.

Case Study

Local underdeterminism and the Younger Dryas event

The climate transition from the last glacial to our current interglacial was not a smooth one. After initial warming from *c*.14,700 years Before Present (AD 1950; hereafter yr BP), an abrupt climate reversal occurred which lasted over 1,000 years. The stratotype for the last glacial-interglacial transition (Termination I) is the GRIP Greenland ice-core (Björck *et al.*, 1998), in which the abrupt climate reversal has been dated from 11,500 to 12,650 GRIP yr BP. This event, Greenland (isotope) Stadial 1 (GS-1), is more widely known as the Younger Dryas. The Younger Dryas was a major climatic event. British mean annual temperatures fell to between −2°C and −5°C, with temperatures of the coldest winter months reaching between −15°C and −20°C (Atkinson *et al.*, 1987). Cooling during the Younger Dryas often resulted in renewed glacial activity in mountain areas throughout Europe (e.g. Benn, 1997). Scrub tundra replaced boreal shrub and woodland communities in northern and western Europe, whilst steppe vegetation replaced woodland communities in many areas further south (Huntley and Birks, 1983). Major declines in human populations or dramatic reorganisations of settlement also occurred across most parts of the northern hemisphere at the onset of Younger Dryas cooling (Anderson *et al.*, 2011). The Younger Dryas influenced other parts of the globe indirectly. For example, Asian summer monsoons weakened (Wang *et al.*, 2001), perhaps due to more persistent winter snow cover on the Eurasian landmass. This would result in slower warming of the Asian landmass during subsequent summers, thereby lowering the land-ocean temperature gradient (Denton *et al.*, 2005), which primarily dictates summer monsoon intensity. There is also evidence to suggest that the influence of the Younger Dryas reached parts of the southern hemisphere, and so was arguably a global climate event (Alley and Clark, 1999). The end of the Younger Dryas event was extremely abrupt; a warming of 7°C in South Greenland was completed within 50 years (Dansgaard *et al.*, 1989). This discovery impressed on the science community for the first time that rapid climate warming can be achieved naturally under surprisingly short timescales.

A change in the strength of the North Atlantic thermohaline circulation (THC) is widely acknowledged to be a causal factor of the Younger Dryas event (Broecker *et al.*, 1985). Differences in water density arising from spatial variability in water temperatures and salinity drive a complex of ocean currents around the planet. The transport of water vapour through the atmosphere from the Atlantic Basin to the Pacific Basin results in Atlantic Ocean surface waters being on average 1‰ higher in salinity than their Pacific counterparts, and this difference is greater at higher latitudes (Broecker, 1992). North Atlantic Ocean surface water density also increases further at higher latitudes through cooling from contact with winter air masses from the Canadian Arctic. This increase in surface water density is sufficient to result in sinking, which leads to the formation of North Atlantic Deep Water (NADW) in the Greenland-Iceland-Norwegian Sea. This high-salinity water mass fills most of the deep Atlantic and flows southward towards Antarctica. The sinking North Atlantic surface water must be balanced by a compensating inflow of water from the south. The major supplier of water to the region of NADW formation is the Gulf Stream. The warm, saline, waters of the Gulf Stream, and its continuation, the North Atlantic Drift, flow northwards, where they densify through cooling and sink; thus NADW is continually produced. The behaviour and operation of the North Atlantic THC has enormous climatic implications. The flux of warm water northwards transfers a significant amount of sensible heat to the atmosphere and supplements annual insolation by approximately 25 per cent in the North Atlantic region (Broecker and Denton, 1990).

The heat released from the ocean as a consequence of NADW formation is carried by the prevailing westerly winds over the European landmass and directly influences the climate north of the Alps and Pyrenees (Rind *et al.*, 1986). The operation of the North Atlantic THC accounts for the mild European winters. Any reduction in the strength of the North Atlantic THC would have direct consequences for climate in the North Atlantic region, and indirect consequences outside of this region (Vellinga and Wood, 2002).

There is broad agreement that the Younger Dryas resulted from a major reduction in the strength of the North Atlantic THC (Broecker *et al.*, 1985). The prevailing view involves an influx of significant quantities of fresh water into the North Atlantic leading to a reduction in surface water density and a significant weakening of the THC (Broecker *et al.*, 1989). The North American Laurentide ice sheet, centred on east-central Canada, was the largest of the former Northern Hemisphere ice sheets (ice volumes of 19.7×10^6 km^3 and a typical thickness of 2,000–2,500 m have been reconstructed by Clark *et al.*, 1996). Melting of this great ice sheet, initiated during the preceding warm event (the Bølling/Allerød interstadial; episode GI-1 in the GRIP Greenland ice core stratotype [12,650–14,700 GRIP yr BP; Björck *et al.*, 1998]) led to the formation of Lake Agassiz, western Canada, which was the largest of the proglacial lakes in North America. At its maximum size, Lake Agassiz was >200,000 km^2 in area and at least 100 m deep (Teller, 1987). Outburst flooding from Lake Agassiz would have increased the flux of fresh water into the North Atlantic, potentially reducing the strength of the THC significantly. Although geomorphological evidence for the flood hypothesis remained elusive for a long time (Broecker, 2006), more recently Murton *et al.* (2010) discovered evidence for an outburst flood path from Lake Agassiz to the Arctic Ocean via the Mackenzie River system. The influx of fresh water into the Arctic Ocean would have ultimately entered the North Atlantic and triggered a major reduction in the strength of the THC.

Several alternative hypotheses have been proposed to account for a reduced North Atlantic THC. For example, Bradley and England (2008) suggest that existing exceptionally thick Arctic Ocean sea ice was the critical source of fresh water to the North Atlantic, supplemented by freshwater drainage from the Laurentide Ice Sheet and Siberian river systems. Firestone *et al.* (2007) propose that one or more large, low-density, extraterrestrial (ET) objects exploding over northern America destabilised the Laurentide Ice Sheet leading to an increase in freshwater flux to the North Atlantic and Arctic Oceans. This ET impact event hypothesis is based on the discovery of numerous carbon-rich layers in sites across North America that, the authors contend, contain impact markers such as iridium, glass-like carbon containing nanodiamonds, and fullerenes with ET helium. These layers date to *c.*12,900 BP (coinciding with the onset of Younger Dryas cooling); the collective layers are referred to as the Younger Dryas Boundary (YDB) by Firestone *et al.* (2007). Surovell *et al.* (2009) refute the ET impact hypothesis after re-analysis of the putative 'YDB' failed to replicate the results of Firestone *et al.* (2007). More recently, Israde-Alcántara *et al.* (2012a) found a 'YDB' layer in the sediment deposits of Lake Cuitzeo in central Mexico, also dating to 12,900 BP. After considering the multiple hypotheses proposed to explain the YDB markers, and addressing the issues raised by the detractors of the impact hypothesis, including those mounted by Surovell *et al.* (2009), they identify a cosmic impact as the only viable hypothesis in explaining the origin of the YDB marker in central Mexico. The evidence for an ET impact and the interpretation of the 'YDB' marker remains controversial (e.g. Blaauw *et al.*, 2012; Gill *et al.*, 2012; Hardiman *et al.*, 2012; response by Israde-Alcántara *et al.*, 2012b).

Rather than invoking catastrophic and unique events, such as an ET impact, or a pulse of fresh water from a proglacial lake outburst as triggers of the Younger Dryas event, Broecker and colleagues have more recently argued that cold reversals equivalent to the Younger Dryas appear to be an integral part of deglaciations (Broecker *et al.*, 2010). They base this argument on a review of the nature and character of the last four glacial-interglacial transitions. Broecker *et al.* (2010) drew attention to the similarities between the characteristics of Termination I and the interval leading up to Termination III (between 240 and 260 ka BP). In particular, Termination III experienced a similar decline in Asian monsoon strength during deglaciation (Cheng *et al.*, 2006), analogous to the weaker Asian monsoon during the Younger Dryas of Termination I. In a similar vein, Renssen *et al.* (2000) view the Younger Dryas as a recent expression of a ~2,500 year quasi-climate cycle but argue for a reduction in solar output as the trigger. In one of the scenarios presented, they propose that, through a series of intermediary processes, a reduction in solar activity would ultimately lead to cooler, but wetter, mid- to high latitudes. An increase in freshwater input to the North Atlantic Ocean, resulting from a combination of increased precipitation in the region and melting of more numerous icebergs, would perturb the North Atlantic THC.

A full and detailed review of the various hypotheses proposed to account for the cause of the Younger Dryas event is beyond the scope of this case study. Nevertheless, the selection of competing hypotheses briefly considered here illustrates well the challenges faced in historical sciences as a result of the phenomenon of local underdetermination. Here we have several plausible explanations to account for the Younger Dryas event. Each explanation is seemingly supported by evidence that is, arguably, more or less strongly empirically equivalent. The problem is further compounded by the fact that some of these hypotheses are incompatible. For example, acceptance of the ET impact hypothesis (atypical event) as the cause of the Younger Dryas would necessarily require the abandonment of, for example, the hypothesis that Younger Dryas-type climate reversals are not untypical of glacial-interglacial transitions. This would then require new questions to be answered as to the cause of a Younger Dryas-type reversal during Termination III. Of course we could accept both of these hypotheses, but the coincidence of an ET impact event at the start of a brief, naturally re-occurring, climate reversal is intuitively difficult to accept. This currently leaves palaeo-scientists in the difficult position of being unable to identify a cause of the Younger Dryas event.

This section has shown that the philosophical issues involved in establishing the nature of 'field' science are varied. A few key points merge however. In studying the field, physical geographers are seeing reality through the lens of critical realism and focus on entities and phenomena identifiable and measurable at the level of research in which these entities interact. The search for causation and explanation is grounded in tendencies and capacities and is concerned with unstable enablers and external validity in specific contexts. Changing temporal scales and spatial locations illuminates the differential operation of these tendencies and capacities in different contexts and so emphasises the potential context-dependent nature of understanding. Finally, local underdetermination means that researchers will often be left with competing hypotheses derived from different underlying theories in their search for explanation.

'Simplifying the field' – laboratory and experimental research

Although it is easy to jump to the conclusion that 'the field' refers to an external, 'natural' environment, the multiple places in which research is carried out makes any simplistic definition difficult. Exploring the relationships that produce phenomena can be carried out in the more controlled and confined setting of the laboratory. Within this setting, researchers are following the 'how to make new evidence' thread of experimental science outlined by Turner (2005) above, although as Cleland (2002) notes, most scientific work is a mix of the two approaches. It could be argued that there is a continuum in the settings for such explorations, from the supposedly controlled laboratory environment, through field experimentation to monitoring the environment as it is. Each step along the continuum implies a decrease in the control exerted by human thought and a corresponding increase in the appearance of reality as it really is in the information gathered. Both views are caricatures and this section will discuss how the illusion of increasingly 'real' behaviour and decreasing human interference hide a set of strategies for understanding 'the field' that still require human thought and, often, human intervention to obtain relevant information.

Rouse (1987, 2008) describes the laboratory as a 'microworld' that he defines as:

> systems of objects constructed under known circumstances and isolated from other influences so that they can be manipulated and kept track of, allowing scientists to] circumvent the complexity [with which the world more typically confronts us] by constructing artificially simplified 'worlds'.
>
> (Rouse, 1987, p. 101)

Laboratories provide the opportunity to produce novel arrangements of some aspects of the world but they are not the world itself, instead they are what Rouse describes as scientific or laboratory fictions. For Rouse, experiments are novel rearrangements of the world that allow the research to assess some aspects of that world that would not normally be as clear. From the flashes of clarity about reality provided by the experiment the researcher can use that information to develop and refine their theoretical or conceptual understanding of a phenomenon. Experiments not only confirm theories, they can prove pivotal in developing them. Altering the parameters of an experiment can help to view the phenomenon in different aspects of its behaviour and so help to extend existing understanding. There is, however, always the risk that the observed behaviour will only ever be found in the 'microworlds' of the experiments and has no bearing on understanding the phenomenon in reality.

Experimental work in physical geography has a long tradition. In rock weathering, for example, there are experiments to investigate salt weathering dating from the nineteenth century (Merrill, 1896; Joly, 1902), and in stone durability dating from the same period, with both traditions extending to the present day (Torok and Prikryl, 2010; Dragovich and Egan, 2011). The early development of an experimental approach and its continuation implies that it is seen as an important route to understanding phenomena in the 'real' world. The central importance of experimentation is in the simplification of reality it seems to represent. Experiments take what the researcher considers to be key parameters or factors in determining a phenomenon and allows the researcher to investigate just those factors or factor in isolation.

Any simplification of reality requires the researcher to select the factors that they consider to be important and then the attributes and attribute values of those factors that they are going to use in the experiment. This selection process tends to be guided by the theories that the researcher believes determine the phenomenon in reality. This belief may be based on observations, traditional practices or on exploration of underlying theories. The selection will also be based on what aspects of the phenomenon the researcher is trying to simulate. Although each subject area is different, the initial experiments exploring a phenomenon tend to try to maximise the effectiveness of the actions of the factors that are varied in the experiment.

For some purposes this type of experimental work is sufficient. In stone durability testing, for example, the sodium sulphate durability test developed at the Building Research Establishment subjects standard-size stone cubes to 10 cycles of 1 hour immersion in 1M sodium sulphate solution and then a fixed period of heating and drying. Weight change from these tests is meant to simulate, in an accelerated manner, the response of the stone to decay processes in the field (Ross and Butlin, 1989). This cost-effective testing of stone durability has economic implications for the quarries whose stone does not fare well in this experimental regime. The exact details of the testing procedure have been scrutinised within the trade (Sedman and Stanley, 1990a, b; Ross and Massey, 1990) and academic literature (Moh'd *et al.*, 1996). The enrolment of this test procedure as the determinant of durability could be viewed as objective, in that it simulates the response of the stone to erosional processes, but Inkpen *et al.* (2004) suggest that the enrolment has much to do with economic needs in the 1920s and 1930s for rapidly characterising building-stone behaviour. Similarly, the test procedure itself was developed within a specific historical context that affected the type of reference stone selected, the technology available for testing, as well as the institutional context that guided what were deemed to be acceptable results from each test. Even for similarly uncomplicated examination of the effectiveness of factors, there is a whole background or context of influences that negate a simple translation and selection of factors from the field to the laboratory. The purpose of the experiment, and the context, has an overwhelming impact on the processes of factor selection and experimental design.

Once the potential effectiveness of a factor is established, research efforts turn to trying to assess whether this factor operates in 'reality' and, if so, how far the effectiveness of the factor is realised. Altering the controlled conditions of the experiment in a manner that the researcher believes increasingly mirrors reality is the usual method of assessing the 'real' effectiveness of a factor. As more and more variables are altered, always in a controlled and precise manner, then the researcher increasingly believes that they are identifying and quantifying the behaviour of the factor as they would observe it in the field. This procedure could be seen within the context of a critical realist view of reality, that is, searching for tendencies and the contexts that determine the extent to which these tendencies are realised. Translating the results of these experiments to the field requires, however, that there is information from the field with which to compare. The process is cyclical and iterative – the field informs experiments and experiments inform an understanding of the field. It also means that monitoring the field and designing, running and interpreting experiments are intertwined as practices and as means to understanding reality.

Experimental work is not confined to the laboratory. The laboratory is just the extreme and most obvious end of a continuum of constructed controlled environments in which investigations are undertaken. The laboratory is a specific space, enclosed and isolated from the phenomena of investigation. Often the laboratory is spatially remote from the field phenomena under investigation. The laboratory is devoid of everything except what the researcher constructs. It is wholly the researcher's world to construct and control as he or she sees fit.

Designing and running field experiments takes the same act of world creation that the artificial environment of the laboratory requires. Constructing an experiment in the field implies that the researcher is bringing a level of control to the field, somehow marking off their section of the 'field' from the rest of the field. A key question should be 'why?' Why would a researcher want to move from the complete control of the laboratory to a constructed and controlled location in the field? The answer lies in the belief that moving to the field is moving towards reality as it is. The researcher may still construct a reality in the field by controlling the nature of inputs and factors but the movement of location to what is considered the 'real' world somehow produces results that are more 'real' than those derived from the laboratory. For example, in a series of field experiments (Hester et al., 1991a, b, c), specific environmental conditions (e.g. light intensity, nutrient availability, simulated grazing) were altered in the field in order to identify the relative importance of environmental variables to the ecological succession from heather moorland to birch woodland. Such a study would have been impossible to achieve in a laboratory. For instance, in order to investigate the importance of light penetration on understorey vegetation communities, birch stands of different age (17, 28, 37 and 63-years-old) were identified in the field, the incidence of light reaching ground level measured, and the plant communities present under the birch stands of varying maturity identified. This particular field experiment revealed that *Calluna vulgaris* was outcompeted and ultimately replaced initially by *Vaccinium myrtillus* as the degree of canopy light penetration fell dramatically from *c.*77 per cent in the 17-year-old birch stand to *c.*29 per cent in the 28-year-old birch stand.

It could be argued that by moving to the field, the researcher is reducing their control. Although the researcher can still construct an experiment, unlike the laboratory, the environment of the field will vary and produce events beyond the perception and control of the researcher. This means that the experiment is in some senses 'wild'. The location lends the experiment a feel of reality, of authenticity, that is lacking when the phenomenon is transplanted to the artificiality of the laboratory. As in the soil erosion experiments outlined, the researcher does try to create a 'typical' critical or Gallieian style experiment, as you would find in the 'hard' sciences. The problem for field experiments is that there will always be the uncertainty of other factors affecting the area bounded as the experimental site. The advantage of undertaking a field experiment is the closeness to reality that this provides. If the phenomenon or factors of interest still operate when the experiments are transplanted from the laboratory to the field, then the belief the researcher has in that phenomenon existing or those factors being key to its operation increases dramatically. In addition, it may be the case that there are factors that cannot be adequately simulated in the laboratory and so field-based experiments are the only alternative. As with laboratory-based experiments, field experiments are a 'world construction', in the sense that researchers devise a world they understand and can manipulate for some purpose. Field experiments do not test reality, as it is, only reality as confined and constructed by the researcher. Paraphrasing Hyndman (2001), the field does not tell us what is out there, it only tells us what the researcher can narrate.

Monitoring the field

Setting up experiments in the field is not the only way in which researchers use the field. The field is also the source of information on phenomena and their variations. Monitoring the field, in terms of the value and variation of specific variables, is a key aspect of physical geography. Monitoring is taken here to include the monitoring of the 'real-time' variation of factors such as temperature, the extraction of proxies of the environment both past and present such as diatoms and the information derived from large-scale analysis of the environment such as analysis of topography.

Chapter 5 covered probing reality; part of that probing involves monitoring reality. This means that much of the discussion in Chapter 5 is also relevant to monitoring. Monitoring is never carried out without some theory in mind that the researcher wishes to assess. Likewise, monitoring is not carried out without some simplification of reality in mind that guides the researchers in deciding what is of relevance to measure as well as constraining what the researcher believes is measurable. Monitoring, therefore, varies with the phenomena the researcher wants to study, the theory under investigation and the instrumentation available to the researcher. On this latter point, the researcher may have to adapt or even invent instrumentation if they believe that the phenomenon they want to study is as yet unmonitored.

Fieldwork as a reflective and imaginative practice

The field is messy and chaotic. It is an open-ended narrative that is designed and told by the researcher. The narrative may be structured and constrained by experience and training but it is a narrative nevertheless. That other researchers can construct a similar narrative using the same methods could be testament to the mirroring of narrative to reality, or the mirroring of narrative to narrative. Whichever it is (and there is no way of knowing which), the act of designing and developing the narrative is a creative and imaginative one. The last sentence is not what you might expect in a text on physical geography, but the role of imagination in deciphering reality in physical geography is one that should be acknowledged and studied.

The dialogue metaphor of study outlined in Chapter 5 requires an active imagination on the part of the researcher to develop and frame questions that they believe reality is capable of answering. Venturing into the field to tackle reality in the raw, an act that this chapter suggests is never possible, again requires an imagination, and an active one at that, to manipulate equipment to capture phenomena as they pass. Baker and Twidale (1991) discuss the re-enchantment of geomorphology, the thrill of the act of discovery, and note that early pioneers in the subject had more chance of experiencing this as everything was unfamiliar. The great imperial projects of the British Empire and the acts of exploration of a hostile interior by Powell, for example, all required acts of imagination to allow interpretation of the strange landscapes encountered. The next stage of subjecting these visual creations to probing by instrumentation required further acts of thought, of imagination, to extract what was considered relevant information. The new worlds required careful, systematic and above all imaginative analysis. Tooth (2006) suggests that even contemporary geomorphology has a similar imaginative encounter as modelling produces virtual worlds to explore.

Frodeman (1995) views geology as a hermeneutic or interpretative science as well as an historical one. By this he means that a geologist assigns different values to what they observe in the field. Different aspects of the field weigh differently or have differing values in their interpretation of the field. Examining the field becomes not just a process of observation but also a process of bringing interpretation to what can be observed. The fieldworker is shifting between observation and interpretation all the time, mingling the two in their search for explanations. In this manner the observed, what we perceive, is always constructed by our conceptions as the two interplay. Conceptions though can be contextually sensitive. Changing context and the conceptions of what counts as evidence, how it fits into theories and what theories are relevant all could change. Exploration of the field becomes a constantly shifting narrative informed by both the field itself and by the observer's conceptions.

Summary

The field is an essential idea for physical geography – it is the external reality with which researchers engage in dialogue. The field as a concept has, however, remained relatively little studied. Fieldwork is an imaginative and creative practice that sits, almost unrecognised, at the heart of physical geography. Some authors have highlighted the potential gendered nature of fieldwork, emphasising the dominating and supposedly masculine nature of the endeavour. Viewing contemporary fieldwork as a more creative and imaginative enterprise can, however, emphasis the collaborative and narrative aspects of the practice. The philosophies of fieldwork relate back to critical realism and abductive reasoning as well as historical science. The search for causation and explanation is grounded in tendencies and capacities and is concerned with unstable enablers and external validity in specific contexts. Underdetermination often means that researchers are left with more than two likely, competing, hypotheses to explain phenomena. Experimentation, whether in the laboratory or in the field, involves creating 'micro-worlds' where phenomena can be isolated. These phenomena are then minutely examined so that variations in their behaviour can be identified and quantified as the parameters of the experiment are slowly altered. The behaviour of phenomena in an experiment may not reflect how the phenomena operate in reality.

Chapter 7

Systems – the framework for physical geography?

Systems analysis in physical geography

Systems analysis came to the fore in physical geography in the 1960s and found formal expression in the classic textbook of Chorley and Kennedy (1971). The introduction of systems analysis into physical geography was not without its critics, but the new approach rapidly became one of the cornerstones of thinking about the physical environment. The continued success of systems analysis is probably best illustrated by a perusal of any set of modern textbooks on introductory physical geography. Often the titles of such texts use 'systems' as an explicit indication of their approach. Almost invariably the contents pages divide the subject matter into specific environmental systems, each of which is considered in turn: for example, the atmospheric system, the lithospheric system and the biosphere, and occasionally there is a chapter attempting to integrate the disparate systems at the end of the text. Systems analysis, if judged by column inches of text in undergraduate books and even research papers, has become the overarching framework for understanding in physical geography. With such success, it is important to understand what systems analysis is. Systems analysis has also moulded thinking about the physical environment. Entities and relations are viewed in a specific framework and studied according to expected modes of system behaviour. An understanding of the constraints on thinking imposed by systems analysis is essential to ensure that the limitations of the systems analysis framework do not become a barrier to comprehension of the physical environment.

Systems analysis did not develop in physical geography; its pedigree is far longer. Systems as a term has been around since modern science developed in the seventeenth century. The rise of systems analysis owes a great deal to the attempt to develop an integrated and all-encompassing framework for all science in the twentieth century. The existence of such a framework implies, first, that all reality is capable of being understood – there are no areas or topics outside of its analytical scope. Second, all reality can be understood in a common framework using the same sets of terms. This means that understanding in supposedly different subject areas does not require specialist terms or specialist knowledge, rather it requires translation of these terms to the common terminology of systems analysis. Third, as there is a common framework, all reality can be expected to behave as predicted by this framework. All reality becomes potentially predictable and, by implication, potentially controllable. Systems analysis should not, however, be viewed as something that emerged fully formed for incorporation into physical geography. Similarly, systems analysis has not remained a static form of study in the 40 or so years it has been used within physical geography. Having said this, the basic tenets of systems analysis remain pretty much the same as they did 40 years ago and the implications of the approach for how the physical environment is studied have become engrained within geographical practice.

Van Bertalanffy proposed general systems theory as a unifying framework in the late 1930s. Tansley pushed the concept of the ecosystem in the same period. Both had a similar view of a system as an integrating concept unifying the different entities found in an environment within a common analytical framework. The advent of cybernetics in the 1940s and 1950s added another level of understanding to the systems approach. Cybernetics is an underrated influence on the development of systems analysis within physical geography. Physical geography tends to focus on its perceived antecedents within ecology, biology and physics as the source of its development of systems analysis. As such, these subject areas hark back to the pre-war era of geographic research. The work of Gilbert (1877, 1896) or Penck (1924), for example, is reinterpreted in systems terminology. Gilbert's use of terms such as system, and his systematic approach to analysis, is taken as an indication of systems thinking even if the terminology had not been invented. Likewise, Penck's view of landscape development is often recast as a systems diagram even though the term was never used by him. Both examples reflect, as noted in Chapter 1, a tendency to reinterpret the past in the light of current thinking. The influence of these individuals and subject areas is not denied, but the nature of the systems they suggested was not the same as the systems as used and understood today.

Cybernetics, the study of self-regulating mechanisms in technology, along with the development of information theory, had a profound influence on the nature of systems analysis that physical geography encountered in the 1960s. Shannon and Weaver's (1949) development of a mathematical basis to information flow and interpretation provided a means of quantifying change in abstract phenomena. Coupled with the lexicon for describing relationships between entities provided by cybernetics, these new analytical frameworks resulted in the development of a highly mathematical and formalised description of systems analysis by the 1960s. It was this version of systems analysis that physical geography tried to bring into the heart of the subject. The success of systems thinking cannot be denied within physical geography, but the total incorporation of a fully-fledged systems analysis is harder to justify.

Systems thinking can be reduced to a few relatively simple ideas, as illustrated in Figure 7.1. The key ingredients of a system are the variables or elements, the relationships between the variables or elements and the bounding of these variables and relationships from the rest of the world. Hall and Fagan's (1956) definition of a system has been used by a number of authors as a starting point (e.g. Hugget, 1980; Thorn, 1988). Their definition is:

> A system is a set of objects together with relationships between the objects and between their attributes.
>
> (Hall and Fagan, 1956, p. 18)

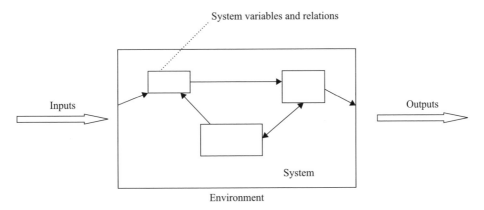

System variables and relations

Inputs

Outputs

System

Environment

Figure 7.1 A simplified system.

The importance of the previous chapters for understanding the limitations of systems thinking should be clearer now. The definition of variables and relationships implies an ability to define and divide the world into distinct entities and relations. Likewise, the definition and bounding or closure of the system itself requires a particular view of reality as divisible and understandable by this division. Systems thinking depends on the reality of the physical environment being displaced from the observer. The observer defines a system made of real entities and relations, the sum of which becomes a sort of super-entity with its own properties and relations to the rest of the physical environment. This super-entity may or may not be simply the sum of its elements and relations. The observer is a passive and objective interpreter of the system, almost by definition outside of the boundaries they have imposed.

A distinction is often made between a system and a model (Thorn, 1988). A system is viewed as an abstraction that is assumed to exist in reality. A model is a fully specified, although abstract and incomplete, version of reality. The distinction is made to clarify the purpose-led construction of models as opposed to the supposed universal nature of systems. A model is designed to serve a purpose, it does not need to fully specify reality, nor to be agreed by all. A system may be unknowable in full, but agreement can be reached that such a set of entities and relationships exists. Adopting this viewpoint may be helpful in distinguishing an operational model from an abstract, but universal, system, but it has the side effect of implying that the system is in some manner more 'real' than the model. The system has some universal status whilst the model does not. Systems, therefore, seem to be a closer representation of reality despite the user-defined nature of all their entities and relationships. Within physical geography it is usually system models (Thorn, 1988) that are being considered rather than systems as universal abstractions.

Application of systems thinking

Development of typologies of systems application has resulted in long lists of different types of systems based on a range of criteria. Chorley and Kennedy (1971) provide one of the first based on the form, function and complexity of the systems studied. Complexity increases from simple cascading and morphological systems to process-response systems and on to biological and social systems. Strahler (1980) develops a typology based on similar criteria to Chorley and Kennedy. Terjung (1976) uses four criteria to separate system modelling in physical geography into different levels. The criteria used relate to the type of logical argument used in explanation (induction or deduction), the level of explanation (individual entities or the system as a whole), the degree of deterministic behaviour, and finally the level of description as opposed to explanation. An important basis for these typologies is the increasing openness of the systems. Isolated systems are an ideal type, ones where there is no movement of matter or energy across system boundaries. Closed systems permit the flow of energy across boundaries, but not matter. Open systems permit the flow of both energy and matter across their boundaries. These distinctions are important as they begin to define the expected behaviour of systems. The definitions derive from physics where an isolated system will tend towards an equalisation of the distribution of energy within it and hence eventually exhibit maximum entropy and disorder or randomness in the organisation of its components. Systems that have their boundaries open to flows, particularly if these flows are of both energy and matter, are able to stave off this 'entropy death' as they retain their organisational structure. Open systems are viewed as able to retain both entities and relationships by maintaining gradients of energy levels between different system components. Flows, from high to low energy, are maintained and so the entities and relationships, the network, that produces the system is maintained. Energy and matter are derived from beyond the boundaries of the system to maintain that system. Although the overall entropy within the universe may be heading towards a maximum level, the smaller system being studied is able to reverse this trend within it by importing energy and matter to maintain its order. Entropy is kept at bay by continually exporting disorder and 'borrowing' energy to keep order.

Each of the above classifications implies an increasingly structurally complex view of the world using systems. Within each classification scheme, the simplest level is viewed as description and definition. From this level the structure of the system is identified (the morphology of the system) to which flows of matter and energy are added (the cascading system). Resultant from these two levels is the process-response system, where energy and matter flow through a set of entities arranged in a specific manner. The flows interact with the entities to produce change or stability in these entities and their relationships to each other.

What complexity means, however, is rarely made explicit. There seems to be an implicit acceptance that the lowest level in the hierarchy represents 'mere' description. At this level it seems to be assumed that there is no real explanation. Setting up the system, defining the entities and relationships to be modelled, appears to be viewed as a relatively simple and uncomplex task. This is the level upon which all the other levels depend and which forms the basis for 'real' explanation. Complexity seems to be associated with the increasing refinement in the specification of entities and relationships. Likewise, a more complex system is one where the entities and relationships are dynamic in the sense of being specified and identified and measured as they change. Full understanding of the system implies that explanation can be generalised and applied to all systems with the same entities and relationships. The problem lies in the fact that full specification would define the uniqueness of the system and so the uniqueness of its explanation is not considered as an issue.

One of the most important impacts of systems thinking in physical geography has been the framework it provides for directing and organising thinking about reality. Thinking is not directed at an individual entity, but at the relationship of that entity to other entities, and the context of the entity within a system. Not only is thinking directed, but it can also be represented in a formal manner (Figure 7.2). Representing reality by symbols that can be applied to different systems provides a strong unifying bond between different parts of the subject. This gives the impression of a uniformity of approach and purpose that, although it may be illusionary, provides a myth of commonality that is lacking in human geography. It also could be interpreted as a singularity of approach and therefore misidentified as a lack of conceptual and philosophical breadth and depth by human geographers. Symbolic representation of diverse systems also means that they can be modelled conceptually and mathematically, and so general relationships can be established and tested between systems. In this manner, trends and patterns identified in one part of reality can quickly be transferred and assessed in another part of reality. Systems analysis, from this viewpoint, has helped to identify the holistic and universal nature of physical reality more rapidly and accurately than any technique before it.

Systems thinking and application in physical geography has not been without problems however. Some of the most important issues relate to the points raised in previous chapters concerning reality and how it is viewed through different philosophies. Central to the problems of systems thinking is the potential to believe that the system is reality. A system is only a model of reality. It is a simplified representation of what the researcher believes to be real and important for the operation of the particular small area of reality they are concerned with. Any system, its components, its relationships and behaviour can only be identified and understood in the context of the theory or theories that informed its construction. Identifying and maintaining the link between theory and the system constructed should be at the forefront of the mind of any researcher. Without this link the rationale for system construction and the expectations for system behaviour are unknowable. Assuming that the system is reality could result in the same system framework being applied inappropriately to different parts of reality.

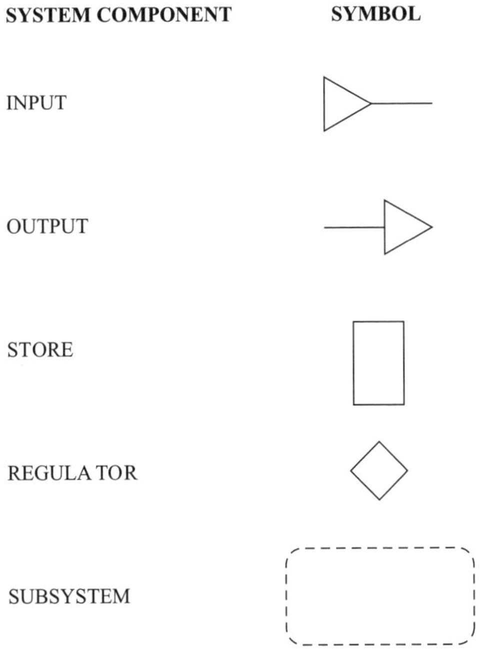

Figure 7.2 Formal representation of system components.

Systems and change

Within the application of systems thinking in physical geography, an important property that has been developed is the ability to identify and predict certain types of behaviour. In particular, the ability to identify when and why systems change or remain stable has become a focus of study. Within systems analysis the types of changes expected are related to how the system and its components change as the inputs change. As inputs, usually considered as discrete events, change in their magnitude or frequency, or both, it is expected that the output from the system or even the internal organisation of the system will alter. The type of change will depend upon how the system is organised and the relationships between variables. Importantly, the presence of feedbacks within the system will determine system behaviour. Negative feedbacks will tend to dampen the impact or tendency to initiate change in system behaviour. Positive feedbacks will tend to enhance changes in system behaviour. Implicit within most assessments of system behaviour is the assumption that the system will behave in a specific and predictable manner. This usually means that the system will exhibit some form of equilibrium behaviour. Equilibrium itself is a complicated and increasingly contentious concept within physical geography and a full discussion of it is left until the next chapter. Chorley and Kennedy (1971) view equilibrium as the maintaining of some kind of balance within a system, whether in association with the relationships amongst the variables, the level of the output, or some steadily moving set of conditions. This highlights the idea of equilibrium equating to balance and being the expected or 'normal' behaviour of the system.

The idea and significance of change in the physical environment did not start with systems analysis, but the framework provided by systems thinking has been useful for formalising the different ideas about change. To clarify how systems analysis has helped to codify change and stability, it is useful to review how time has been seen within physical geography and then how change has been incorporated into physical systems.

Change (and stability) can occur in both time and space, although it is temporal change that has been the focus of most study. Discussion of temporal change has been seen as a key ingredient in defining contemporary physical geography. Schumm and Lichty's (1965) paper on 'Time, Space and Causality' is still viewed as a classic presentation of different scales of temporal change, even if the theory is untestable. They identified three types of time: cyclical, graded and steady-state (Figure 1.5). At each temporal scale different variables will be important for the operation of the physical system. As the temporal scale changes what were previously important variables effectively become constant, unchanging variables, and now represent the context of the system. Although Schumm and Lichty were aiming their critical review at the Davisian model of landscape development, their use of concepts from other parts of science, such as equilibrium, in developing their argument provided a theoretical basis for changing the emphasis of what geomorphologists did. They highlighted the need to focus on the appropriate scale of study and the variables appropriate to that scale. Davis was not disproved, merely regarded as irrelevant for more process-orientated studies that focused on reality at a different scale from Davisian landscape evolution. In this respect, although Schumm and

Lichty did not use system terminology, their study highlights the importance of appropriate boundary and entity definition for studying reality. Likewise, the use of diagrams such as that depicted in Figure 7.3 provided a set of expected behaviours for the scale of reality being studied. They had started to define what type of changes should be expected in particular variables under certain conditions. They had started to define systems behaviour in geomorphology.

Wolman and Miller (1960) looked at how the magnitude and frequency of events could be used to understand change and stability within the physical environment. Although they limited their definition and exemplification to fluvial systems in temperate environments, their form of analysis did provide an insight into change. They identified the magnitude of an event by how much 'work' it did in a catchment. Work was defined by the amount of sediment moved by events of given frequency. Importantly, the sediment that had been moved had to be measured as output from the catchment in order to count as having been 'worked on' by an event. A 'normal' event was viewed as bankfull discharge. This definition limited the definition of geomorphic work to a relationship with sediment movement out of the catchment. Movement within the catchment, or work done in weathering material for transportation, was not included, for example. With this definition of work, it is clear that large events do most work, moving most material. Wolman and Miller also noted that events of a given magnitude occur with differing frequencies. Large events are relatively rare, small events are common. Combining the two trends for

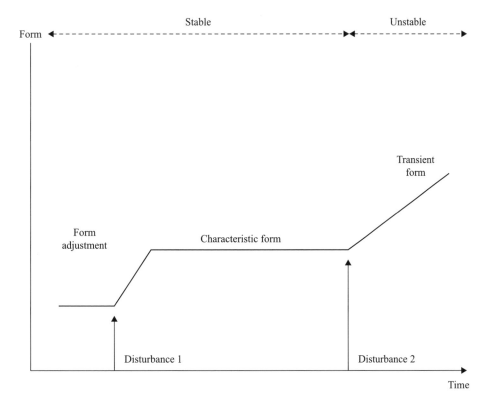

Figure 7.3 Characteristic and transient forms, based on Brunsden and Thornes (1979).

events, they produced the magnitude/frequency curve. From this curve they identified medium-size events as doing most of the work within a catchment. They linked these medium-size events to bankfull discharge within the catchment and identified forms such as river banks as being adjusted to these size events. They were, in other words, defining expected behaviour for the catchment system in terms of river bank morphology and 'normal' events. The relation between form and event was, however, defined only in terms of the absolute magnitude and frequency of the event. The form and its reaction and inter-action to an event is assumed to be constant, the system is devoid of feedbacks. They did, however, draw a distinction between events that may be formative – that may produce landforms in the landscape – and those that did the 'normal' work. In other words, they recognised the potential for qualitatively different types of events in the landscape.

Wolman and Gerson (1978) developed a more complicated version of the relationship. In this paper they define event effectiveness not in absolute terms, but in relation to the ability of the landscape to 'restore' itself. This assumed a tendency for the landscape as a whole to move towards a preferred characteristic or equilibrium state once a disrupting event is removed. The event is still viewed as a discontinuity in the system, as an inter-ruption of the 'proper', normal behaviour of the system. Once the event disappears, the system, in their case the landscape, can once again return to its normal modes of opera-tion. Events are not seen as important parts of how a system operates in these papers, instead they are seen as disruptions to be endured. Event effectiveness did not solely depend upon the property of magnitude. The context of occurrence was vital. If two large events occur one after the other, the second event would have no sediment to remove from the catchment because the first event would have already removed it. Event sequences became important, so the event in the context of other events had to be consid-ered. In addition, an event of a given magnitude would have a differential impact depending on the power of the restorative forces in a catchment. Although Wolman and Gerson tended to limit these forces to vegetative regrowth, the concept could be applied more widely. This new definition highlights that the context within which an event occurs is important for the relationship between event magnitude and frequency and landform change. In other words, the behaviour of the system depends upon the system as much as it does upon the input into the system. The presence of strong negative feedbacks can dampen the changes initiated by an event. Weak negative feedbacks reduce the ability of a system to 'repair' the damage of an event. It might even mean that the system could not return to its previous 'normal' mode of behaviour.

Brunsden and Thornes (1979) added a further dimension to expected system behaviour in their version of landscape sensitivity. They defined landscape sensitivity as:

> Landscape stability is a function of the temporal and spatial distributions of the resisting and disturbing forces and may be described by the landscape change safety factor here considered to be the ratio of the magnitude of barriers to change to the magnitude of the disturbing forces.
>
> (Brunsden and Thornes, 1979, p. 476)

This definition highlights the significance of both temporal and spatial changes in a system due to an event. Events are no longer regarded as simple single inputs, their nature can change and in so doing the behaviour that they induce changes. Likewise, systems are no longer viewed as simply responding to an event, the event is instead mediated through the system and the mediation itself becomes a form of creeping change within the system. Events still have impacts, but the impact is more disparate and complicated than previously envisaged. System behaviour still adheres to the idea of equilibrium, but the nature of change depends more upon the interaction between system state and input rather than just input. In Brunsden and Thornes (1979) different timescales for individual landforms become important. Time can be divided into the time taken to recover from a disruption (relaxation time) and the time period during which the 'normal' or characteristic form is present. The characteristic form is the form the landform takes when it is able to absorb disruptive events. Relaxation time is a time period when a landform is adjusting to a disruption and when it exhibits non-normal forms. These transient forms are not in equilibrium with the processes forming the landform, in other words they are not characteristic forms. Landforms will have different stabilities and react to the same disruptions in different ways depending on their state, that is if they are adjusting, or have adjusted already, to disruptions. Figure 7.4 illustrates this idea by viewing the landform as being entrenched to differing degrees in its current state. Different parts of the landscape will have different degrees of entrenchment and so different sensitivities to the same event. In this way Brunsden and Thornes provide a very complex and spatially and temporally differentiated view of landscape and landform behaviour, but one that retains at heart the idea of a tendency toward some equilibrium condition.

From the above discussion it should be apparent that the significant aspect for the behaviour of the system is the relationship between the resisting and the disrupting forces. These general concepts can only be made operational by the definition of the researchers. There are no hard and fast rules as to how to define each of these properties, but their conceptualisation is vital for understanding how systems behave. Brunsden and Thornes's analysis requires that entities in the system be defined in relation to an input or event as either resisting or enabling its mediation as a disruption to the 'normal' operation or behaviour of the system. In this manner, the definition of normality, or expectations of system behaviour, become essential to defining what to study and how to understand

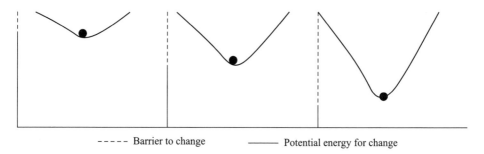

----- Barrier to change   ——— Potential energy for change

Figure 7.4 System entrenchment and relation to landscape stability, based on Brunsden and Thornes (1979). (a) Unstable situation. Height of barrier to change less than potential energy for change. (b) Stability. Height of both barrier to change and potential for change are the same. (c) Entrenchment. Barrier to change much greater than potential energy for change. System highly stable.

changes in system behaviour. Understanding the system is inherent within the manner in which the system is constructed.

Schumm (1979), in the same volume as Brunsden and Thornes, further developed the concept of thresholds in a system and the potentially complex behaviour that even simple thresholds could produce. Thresholds mark points in the system where the behaviour changes from one mode of operation to another. In other words, the system stops following one expected course of behaviour and, after some brief odd behaviour, settles down into a different, but expected (i.e. predictable), course of behaviour. Thresholds can be extrinsic, a system responding to an external influence. Thresholds can also be intrinsic with the switch in behaviour occurring without any change in the value of the variables external to the system. Adjustment of a landslide to the continual process of weathering of its regolith could be viewed as an internal threshold. At some point the continual operation of weathering agents could weaken the strength of the regolith sufficiently to induce a failure during a rainfall event that previously would have passed without incident. Schumm identifies a geomorphic threshold as one that is inherent within the system, one that develops through changes in the morphology of the landform itself, such as suggested above.

A system does not necessarily respond rapidly to a threshold being crossed. Schumm uses the example of an experimental channel where there has been a single change in base level. Incision at the river mouth progresses upstream, rejuvenating tributaries and removing previously deposited sediment. As this erosive wave progresses upstream, the sediment load increases in the main channel with consequent deposition and aggradation in the previously incising channel. Eventually, the tributaries become adjusted to the change in base level and the sediment supply to the main channel dries up. This initiates a new phase of channel incision. From a single event, the change in base level, a series of incised channels and terraces have been formed and different parts of the catchment have responded at different rates and in different ways to the same stimulus. This experiment, plus the example of Douglas Creek that Schumm uses, implies that even a simple system can have a complex response to events. The precise sequence of changes will depend on the context of the system and its thresholds. Importantly, there remains the problem of identifying sensitive parts of a system and the location of thresholds *before* changes occur.

Case Study

Systems and landscape sensitivity

Landscape sensitivity as expounded by Brunsden and Thornes (1979) is an important means of understanding systems and change. In explaining the idea and in developing its application (Brunsden, 1990, 2001), use is made of major concepts in systems thinking. The application of these concepts provides a systems framework that could be applied to most phenomena in the physical environment. From a Popperian perspective this could call into question the basis of landscape sensitivity as a theory, casting it instead as a myth. In Brunsden and Thornes (1979), four fundamental propositions concerning landscape development are put forward as the basis for constructing the idea of landscape sensitivity (Table 7.1). These propositions could be viewed as basic postulates of the theory or as untestable truisms in the sense put forward by Haines-Young and Petch (1986). Brunsden and Thornes (1979) view landforms as either being in equilibrium with environmental conditions, i.e., characteristic forms, or as forms moving towards that state, i.e., transient. The landscape is continually in a state of change as external inputs force it to adjust towards its characteristic state, or as internal thresholds are crossed. The distribution of disruptions, and therefore of landforms in different states of transience or stability, is both spatially and temporally complex. The result is a complex assemblage of landforms in varying stages along a sequence of adjustment. The susceptibility of any specific part of a landscape can be expressed by the relationship between the forces resisting change and those forcing it. In other words, there are spatial and temporal variations in the balance of the landscape and its components. Sensitivity expresses how close the landscape and its components are to the edge of this balance.

Table 7.1 *Propositions for landscape systems, based on Brunsden and Thornes (1979).*

Proposition	Description
Proposition 1	For any given set of environmental conditions, through the operation of a constant set of processes, there will be a tendency over time to produce a set of characteristic forms.
Proposition 2	Geomorphic systems are continually subject to perturbations which may arise from changes in the environmental conditions of the system or from structural instabilities within. These may or may not lead to a marked unsteadiness or transient behaviour of the system over a period of 10^2–10^5 years.
Proposition 3	The response to perturbing displacement away from equilibrium is likely to be temporally and spatially complex and may lead to a considerable diversity of landforms.
Proposition 4	Landscape stability is a function of the temporal and spatial distributions of resisting and disturbing forces and may be described by the landscape change safety factor, here considered to be the ratio of the magnitude of barriers to change to the magnitude of the disturbing forces.

Brunsden (2001) emphasises that landscape sensitivity permits an assessment of the likelihood that a given change in the controls of a system will produce a sensible, recognisable, sustained but complex response (Brunsden, 2001, p. 99). The image of the landscape system is one that is in a state of possible change. Whether change occurs or not depends upon the characteristics of the system but always in relation to the disruptive forces affecting it. The landscape system is balanced between its propensity for change and its absorption of the disruptions within its existing structure. Brunsden (1993, 2001) outlines several sources of resistance to, or absorption of, disruptions (Table 7.2). These usually reflect the structure or morphology of the system under investigation. System response will also depend on the sequence of events experienced. Different sequences will activate different pathways in the system and so create different responses. As the system responds to events, it alters its own characteristics and so what Brunsden calls 'time-dependent preparation processes' can alter the sensitivity of the system. Weathering, for example, can weaken the bonds within a rock mass and increase its susceptibility to failure by future disruptive, triggering

Table 7.2 *System properties and system behaviour, based on Brunsden (2001).*

System properties	Description
Strength resistance	Barrier to change imparted by the properties and dispositions of the materials out of which the system is made.
Morphological resistance	Variable distribution of potential energy across the system. Scale-dependent depending upon the processes considered.
Structural resistance	Design of system – components, topology, links, thresholds and controls. Subdivisions are location resistance (location of system elements relative to that of the processes capable of generating change), transmission resistance (ability of system to transmit impulse of change).
Filter resistance	System control and removal of energy from landscape – shock absorbers of the system.
System state resistance	Ability of system to resist change because of its history. Past history will have configured system pathways in a specific manner, so no two systems will respond the same to the same input.

events. Brunsden refers to the Crozier *et al.* (1990) model of slope-ripening by increasing regolith thickness through weathering as an example of a preparatory process. Operating in the opposite direction, changes in the system can increase its resistance to disruptions. Church *et al.* (1988), for example, suggested that the coarsening of bed surface textures occurs in tandem with a rearrangement of the coarse clasts into increasingly stable geometric arrangements.

Development of landscape sensitivity has resulted in a vast number of related ideas that emerge as logical extensions of the initial propositions. The couching of each in systems terminology provides a means of incorporating these into systems thinking. The range of ideas, however, is almost too inclusive. There are not many types of behaviour observed within geomorphology that could not be defined under the descriptive preview of one of these ideas. Testing the presence of a specific type of change is difficult. In the case of preparatory processes, they can only be identified after the change; before the change they are only one of many possible processes of change. Similarly, resistances as identified in Table 7.2 are defined by the nature of the system, an entity identified and populated with elements by the observer. System resistance is built by the knowledge and experience of the observer, but it is unknowable whether there is a basis or even correspondence for this in objective reality.

Summary

Systems analysis has been a dominant method of thinking about the physical environment since at least the late 1960s. Systems analysis provides a set of formal and standard terms for translating the physical environment and a common framework for analysing the whole of the physical environment. Systems are still, however, a simplification of reality, not reality as it really is. Despite some discussion of the meaning of the term, systems thinking has been applied extensively. Systems thinking has, however, constrained the type of behaviour researchers expect from the physical environment. Specifically, features such as positive and negative feedbacks and equilibrium are assumed to be present and to be 'explanations' for behaviour. Likewise, there has been a tendency to use and apply systems terms, such as robustness, sensitivity and relaxation to the physical environment. These terms have very flexible definitions depending upon the context in which they are applied. This makes the testing and falsifying of theories expressible in these terms problematic, as many translations of these terms are possible.

Chapter 8

Change and complexity

Equilibrium – an ex-concept?

Understanding change in the physical environment has meant understanding and applying the concept of equilibrium. Equilibrium and its validity is the focus of a continuing debate amongst geomorphologists (e.g. Mayer, 1992; Welford and Thorn, 1994). Despite the disquiet with the notion, it still has a powerful hold over how physical geographers think about reality. Even where alternative concepts are offered, they often refer to equilibrium to illustrate their difference from it. Equilibrium acts as a reference point for debate, even for its opponents. Given the important position of equilibrium, it is vital to understand what it is, how it has been applied, and how it has affected how physical geographers think.

Equilibrium is a concept borrowed, as usual with physical geography, from the 'hard' sciences of physics and chemistry. As with most appropriated concepts, its application in physical geography has not adhered to the stringent definitions applied in these subjects (assuming they ever were applied stringently in the first place). For example, within statistical mechanics, equilibrium is viewed as the most probable macrostate of a system composed of potentially different microstates (Welford and Thorn, 1994, p. 670). Howard (1988) makes the point that the definition of equilibrium relies on a great deal of subjectivity in the sense that the researcher often defines the input and output variables in a search for equilibrium. Similarly, the variables selected for measurement must be capable of changing over the time and space scales of the study. Likewise, Howard (1988) also emphasises that just using the term equilibrium does not make a statement about cause and effect within a system. Instead it describes a presumed state without necessarily explaining that state.

Renwick (1992) introduces the term disequilibrium to describe landforms that tend toward equilibrium but have not had sufficient time to reach this condition. This is contrasted with non-equilibrium landforms that can undergo rapid and substantial changes in form such that there is no average condition to which they seem to tend over time. In this manner Renwick distinguishes two types of opposites to equilibrium rather than a single opposite.

Equilibrium implies some sort of balance as well as the maintenance of that balance. In other words, equilibrium implies both a condition for a system and a behaviour for the system in order to maintain that condition. Within physical geography, this idea of balance has been applied at a variety of scales. It has been used to refer to the overall state of the system, as well as to describe the status of individual variables. Some researchers have claimed that equilibrium is a term that should only be applied to the last of these states. More contentious has been the extension of equilibrium to apply to systems that are changing in their states.

Equilibrium is also used to describe trajectories of system behaviour (Figure 8.1). Within a systems framework the interaction of variables will produce changes in both the variables and the system itself. If measured outputs from the system are taken to represent its behaviour, then these outputs for any given time period can either remain the same, increase or decrease. Plotting these changes over time indicates the trajectory of system behaviour. A similar set of changes can occur at the level of individual variables or between variables. Schumm and Lichty (1965) make use of this expanded definition of equilibrium in their analysis of time and causality in physical geography. This view of equilibrium has spawned a range of different types of equilibrium such as metastable, dynamic and quasi-dynamic. Interestingly, there is even non-equilibrium, using equilibrium behaviour to define alternative behaviour by its absence. Each of these definitions is

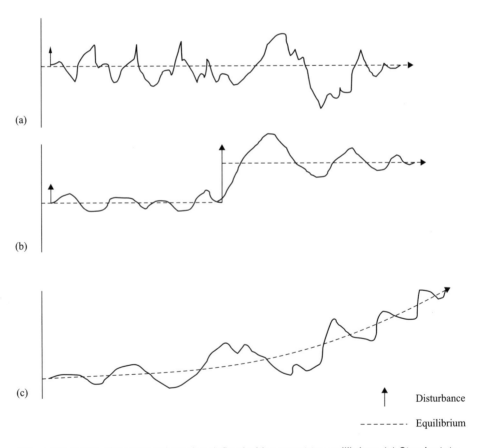

Figure 8.1 Types of system trajectories defined with respect to equilibrium. (a) Steady-state equilibrium. (b) Metastable equilibrium. (c) Dynamic equilibrium. Combining these equilibria can produce a range of different system behaviours.

possible to apply, because equilibrium is viewed as the typical behaviour of a system but without any detailed explanation of why this is so.

Using a systems framework implies that it may be possible to identify feedbacks and thresholds within a system and, importantly, these will influence how the system behaves. Equilibrium relies on the presence of negative feedback loops within the system. This means that researchers begin to define an expected behavioural pattern for systems. Inputs may vary, but the system will respond by tending to dampen the impact of these changes and returning to its previous condition. The system will have a 'normal' behaviour and tend to want to return to it. This expectation raises issues of teleology in system functioning, i.e., that the system behaves as if it had a predefined function. These types of behaviour are generally explained away in terms of system tendencies rather than imputing functional behaviour to a system. The expectation of an identifiable 'normal' state for any physical system does imply that change is unusual in reality. Researchers will therefore tend to look for evidence of change being reduced rather than change being amplified. The latter suggests positive feedback, which will create unstable structures within systems. Stability of entities and what appear to be coherent systems to the researcher imply that reality must be dominated by negative feedbacks.

By disturbance, researchers have tended to mean events or rather inputs into the system. A range of different types of inputs have been identified (e.g. ramped, pulsed, continuous), which can all influence system behaviour in different ways and over different timescales. If system equilibrium were defined as a fixed relationship between input and output, or between variables and their states, then the concept would have limited application in reality. Inputs vary both as discrete units and as continuous values and as a result so do outputs. There are fluctuations in both and hence in the variable states that produce them. Ironing out these variations has been the basis for identifying equilibrium as an average state of a system. Once an average is involved, the idea of equilibrium as a user-defined state begins to become clearer. An average implies the selection of a timescale (and more rarely a spatial scale) over which values for the system state can be averaged. Likewise, it also implies a series of measurable system states at a temporal resolution (or spatial resolution) shorter than the averaging time period. In other words, as noted in previous chapters, the researcher makes the world, the reality, of the equilibrium via their measurement systems and theoretical constructs. Equilibrium becomes identifiable because the researcher is trying to find it.

Defining equilibrium by an average system state provides the basis for defining steady-state equilibrium. It also, however, permits the extension of the concept to situations where the average state is not constant. Although inputs and outputs may change, the structure of the system may remain constant. The morphology of the system may remain constant even if the flows through it increase or decrease. In this sense, the system can be said to be stable. Researchers have tried to capture this stability of morphology, as opposed to constancy of input and output values, by the development of concepts such as dynamic equilibrium. Dynamic equilibrium refers to the progressive change of a system around an average fluctuating state. This type of system behaviour appears to show the system changing and indeed it is. The system structure, how flows are organised and how disturbances are nullified, remains constant. What alters is the value of variable states within the system. Within a slope system undergoing slope decline, for example, the manner in which material is moved across the slope surface, the transfer of energy from rainfall input to movement of soil material, does not alter. What alters is the rate of this movement. Flows alter because state variables alter; in the case of this example, slope angle slowly declines and so other connected variables such as slope length and height alter. The state of system variables is therefore undergoing constant and predictable change. In no manner could this be described as a steady-state equilibrium even in terms of an average state. This does not, however, capture the point that the operation of the system remains the same. Dynamic equilibrium allows this point to be conceptualised. It highlights that change can occur in the system but that stability in system functioning is preserved and prevails. Once again, the idea of a system tending towards a normal mode of operation, and that this stability represents the normal state of the physical environment, is central to systems thinking.

Preservation of system morphology is also a theme within the development of definitions of equilibrium such as metastable equilibrium and quasi-equilibrium. On occasions, systems may change their behaviour dramatically. Schumm's (1979) identification of internal thresholds within systems implies that progressive change may result in the crossing of an irreversible boundary. A point of no return is reached where the structure of the system breaks down or begins to operate in a different manner. At this point the previous expected types of system behaviour disappear. They are replaced by a new 'mode' of behaviour: a new normal condition is established for the system. The flip between states is the metastable part. The system approaches a threshold in either its overall behaviour or for individual variables. Crossing this threshold changes the relationships between variables and a 'new' system state is born. For example, in applying the systems framework to observations of coastal ecosystem change over time, Shennan (1995) provides the example of a rise in the water table resulting in the crossing of a system threshold and a consequent change in system state. The crossing of the system threshold results in the replacement of a fen carr community by a reedswamp community – the rising water table means that conditions are too wet to support fen carr communities.

The identification of this change is up to the researcher. It may involve the redefinition of the system structure to include new variables or the redefinition of how variables interact or what they are. The important point is that there is a user-defined restructuring of the system. Despite this restructuring, there is still an expectation that the 'new' system will continue to attempt to establish and maintain an equilibrium, that a 'normal' set of behaviours will emerge and function. At this point it may be possible to begin to define domains of expected behaviour. The possible trajectories for a system can be mapped out in terms of this behaviour space. This means that researchers will expect system behaviour to focus about these 'norms'.

Of equal significance, researchers will predict what changes will result in the type of equilibria being reached and how long a system might occupy a specific equilibrium. In other words, the physical environment becomes a predictable entity. Brunsden and Thornes' (1979) use of 'characteristic' forms to represent landforms in equilibrium with their environmental conditions, and 'transient' forms as landforms tending towards characteristic forms and so not in equilibrium with their environment, are examples of the pervasive grip of equilibrium. Characteristic forms are viewed as the 'normal' forms, whilst transient forms are ephemeral, unexpected and not 'normal'. This underlying set of expectations is further enhanced by the use of both terms in defining landscape stability – the ratio of characteristic as opposed to transient forms; in other words, the ratio of normality to oddities in the landscape. The greater the oddities, the less stable the landscape.

The expansion of the range of definitions of equilibrium means that the opposite of what equilibrium is becomes harder and harder to define. The flexibility, fuzziness and adaptability of definitions of equilibrium are a key strength of the concept and also its key weakness. Some have claimed, and not without cause, that equilibrium can mean anything and so is of little use as a workable and explanatory concept for understanding the physical environment. As the sphere of system behaviour encompassed by the term equilibrium increases, it becomes increasingly difficult to determine any sort of behaviour that cannot be redefined as some sort of equilibrium. Mayer's (1992) definition of non-equilibrium forms still uses the standard system behaviour diagrams to illustrate how unlike equilibrium his non-equilibrium forms are. Such problems of trying to define an alternative to equilibrium illustrate how common and constraining the thinking developed by the concept has become in physical geography. The emergence of chaos theory and complexity in the last 30 or so years has suggested that there may be other ways of thinking about the physical environment and how it changes, but the use of these concepts has shown how engrained equilibrium is within physical geography.

Chaos and complexity – more of the same?

Chaos and complexity are not the same thing. Although Manson (2001) recognises that there are different types of complexity, each with different and often contradictory assumptions, he still attempts to provide a threefold classification of complexity. Algorithmic complexity refers to complexity as the simplest computational algorithm that can describe, and so reproduce, the behaviour of a system (Manson, 2001, p. 406). Deterministic complexity is related to chaos in Manson's typology. This type of complexity displays sensitivity of system trajectories to the initial system conditions. It also assumes the possibility of attractors to which a system trajectory tends in amongst the seemingly chaotic behaviour. It is this form of complexity that Manson feels appeals most to postmodernist human geographers. Aggregate complexity refers to the holistic behaviour resulting from the interaction of system components. These interactions produce emergent behaviours, behaviours that cannot be predicted from the individual components alone. The basis for this emergent behaviour lies in the interaction of system components. Reitsma (2003) provides a critique of Manson (2001). She suggests that Manson's typology is one that identifies different measures or definitions of complexity rather than different types. She also carefully draws a distinction between complicated and complex. A complicated system is one where a complete and accurate description of a system can be given in terms of its individual component parts. Although such a description may contain a lot of information and a lot of data, the operation of the system can be predicted from its component parts. Indeed in the examples Reitsma presents of a computer and videocassette recorder, the predictability of the machinery is the whole point of their construction. In contrast, complexity refers to a system where the description of its components does not provide enough information to predict the behaviour of the system. She suggests that often complexity is assigned to complicated systems. In a reductionist view of the world, complicated systems are likely to be the norm, as parts predict the whole, whilst in a world of complexity, emergence reigns and the whole is unique relative to the parts.

Reitsma also states that chaos and complexity are not presented as clear and distinct theories within Manson's discussion. Reisma accepts that there is no common framework or set of definitions for distinguishing chaos and complexity, but the two have to be clarified as different. Chaos theory is concerned with the operation of simple, deterministic, non-linear, dynamical and, importantly, closed systems. It is these systems that are sensitive to initial conditions and which can produce seemingly chaotic behaviour under the action of slight perturbations. Complexity theory is concerned with complex, non-linear and, in contrast, open systems. Complex systems, rather than 'degenerating' into chaotic behaviour, respond to a perturbation by organising their components into emergent forms that cannot be predicted from the system components themselves. This is system self-organisation.

Table 8.1 outlines the two typologies suggested by Manson (2001) and Reitsma (2003) along with the difference between these two classification schemes. They are similar, however, in what they are trying to say about physical reality. They are both alternatives to what they regard as a restrictive and generally simple, linear view about relationships and change within the physical environment. Most introductions to these concepts

Table 8.1 *Complexity and chaos typologies.*

Manson (2001)	
Algorithmic complexity:	Simplest computational algorithm that can reproduce system behaviour. Complexity lies in difficulty of describing system characteristics mathematically.
Deterministic complexity:	Interaction of variables produces systems that can be prone to sudden discontinuities. Sensitivity to initial conditions and bifurcation are key characteristics.
Aggregate complexity:	Individual components of a system operate to produce complex behaviour.
Reitsma (2003)	
Deterministic complexity:	Based on information theory. Algorithmic content of a string of bits. Complexity equated with randomness.
Statistical complexity:	Measure degree of structure present. Randomness equates to maximum complexity.
Phase transition:	Maximal complexity is mid-point between order and chaos.
Chaos derivatives:	Precise definition through indices such as Lyapunov components – system's sensitivity to initial conditions.
Connectivity:	Complexity is measure of degree of connectivity in system. Greater connectivity equates to greater complexity. Oddly, this may mean greater system stability.
System variability:	Increase in system variability reflects an increase in complexity.
Relative and subjective complexity:	Complexity arises because of human perception and so only exists relative to the observer.

define them as contrasts to the Newtonian view of the world. Although this caricature is shorthand for a particular linear modelling view of the environment, it is also a straw-man that is probably an inaccurate description of how most researchers perceive the physical environment anyway. Nonetheless, the image of physical reality painted by the linear view does sound familiar. Variables interact in a regular manner and we can deter-mine patterns in this regularity that we can model so as to predict future behaviour. An important part of this transferability of prediction is the idea that system states that are close together will evolve or change along similar, if not identical, trajectories. Chaos theory holds that systems may be extremely sensitive to their initial condition and that change may be non-linear rather than linear (Figure 8.2). Both these conditions imply that the system will behave in an unpredictable, if not chaotic, manner. Even if the initial

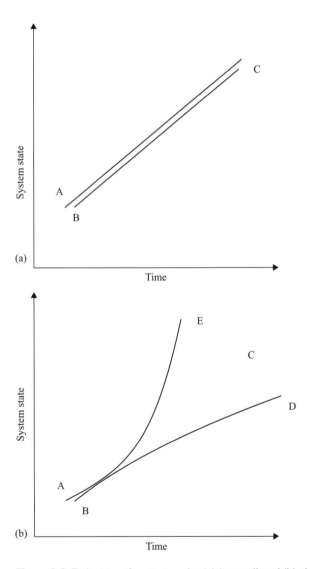

Figure 8.2 Trajectory of system under (a) 'normal' and (b) chaotic behaviour.

state of a system can be measured, the accuracy of that measurement may be insufficient to preclude such behaviour. In this sense chaos theory goes against the grain of scientific thinking since Newton. It begins to suggest that some 'laws' of nature may be unknowable to us because they produce irregular behaviour in our user-defined physical systems.

Phillips (2003) provides a set of common types of complex non-linear dynamics that provide a wide coverage of different types of system behaviour. This implies that most behaviours seen in the field could be interpreted within the framework of chaos and complexity, or as Phillips prefers, non-linear dynamical systems. Phillips (2003) also identifies that non-linearity has a variety of sources within the environment. His definition of why the sources are non-linear, as well as the examples used, highlight the continuing view that geomorphologists have been studying non-linear relationships all along, but have not recognised these as such. Importantly, within Phillips's scheme is the point that the presence of non-linearity does not necessarily mean that chaotic or complex system behaviour will occur as of necessity. Complexity is not relevant to every geomorphic problem, even if the potential for it to occur is present. The sources of non-linearity are varied and have often been explored and explained by other means by geomorphologists. This does not negate the potential importance of non-linearity for providing other types of explanations for complex behaviour, but it means that non-linearity need not be the obvious or immediately clear source of complex behaviour. Non-linearity is only one tool in the explanatory kit of physical geographers.

Within physical geography, chaos and complexity have been mentioned but still not widely embraced. Where they have been seriously assessed, it is usually in relation to existing concepts such as equilibrium. Phillips (1999), for example, provides an outline of some of the definitions used within complex non-linear dynamics. The two concepts, equilibrium and chaos and complexity, seem to produce similar definitions. Phillips's analysis seems to suggest that physical geographers had been using complexity all along but just had not realised it. This view suggests that chaos and complexity theory are already operating within physical geography, so there is little that physical geographers need to do to fully incorporate it. It also implies that the types of explanations offered by chaos and complexity theory are already familiar to physical geographers and so little needs to be done to adapt physical geography to these new ideas. Alternatively, it could suggest that there is little point in physical geographers adopting the new concepts as they add very little to the explanatory frameworks they already have. There may be something to this point, but it may also be fair to say that the explanatory framework suggested by these two theories is one that a lot of physical geographers would feel uncomfortable with.

Case study

Chaos theory and ecological systems

In the classic Kuhnian sense, the development of chaos theory as a view of seeing the world in a different light represented a paradigm shift in science. Chaos refers to a type of system behaviour in which small changes in initial conditions can give rise to disproportionate changes in system behaviour, and thus long-term prediction of system behaviour is impossible. The identification of system-sensitive dependence on initial conditions was first discovered in the field of meteorology by Edward Lorenz (Lorenz, 1963). Whilst attempting to generate computer-based model projections of weather, Lorenz discovered that extremely small (and thus seemingly unimportant) differences in model inputs resulted in unpredictable system behaviour. This explains the limited time horizon of local weather forecasting (Lorenz, 1982). The propensity for chaotic system behaviour is not limited to meteorology, but appears to be pervasive. Behaviour characteristic of chaos has been identified in population biology, fluid dynamics, cosmology, physiology, economics and the social sciences to name but a few (Gleick, 1987).

Fluctuations in species abundances can exhibit stable equilibria, periodic and quasi-periodic cyclical behaviour. It can also exhibit chaotic behaviour, which is characterised by erratic and unpredictable changes in population numbers, with the pattern of variation sensitive to small changes in initial conditions. The discovery that complex chaotic dynamic behaviour can emerge from simple population models (May, 1974, 1976) led to a paradigm shift in ecology, and chaotic behaviour has since been recognised in mathematical models of population dynamics (e.g. Costantino *et al.*, 1997), predator-prey interactions (e.g. Vandermeer, 1993), food-chain dynamics (e.g. Van Nes and Scheffer, 2004) and competition for limiting resources (e.g. Huisman and Weissing, 1999). Nevertheless, despite mathematical models demonstrating the capacity for chaotic behaviour, this has seldom been observed in actual populations.

The identification of chaotic behaviour in population numbers is necessarily restricted to the analysis of organisms whose existence is short-lived, and in which multiple generations can be observed within a time period convenient for research. In one of the first examples of its kind, Beninca *et al.* (2008) isolated a planktonic food web from the Baltic Sea, consisting of two nutrients (nitrogen and phosphorus), detritivores and ten different functional groups comprising bacteria, several species of phytoplankton, herbivores and predatory zooplankton species. The food web was cultured in a laboratory mesocosm and species abundances monitored for over eight years. External conditions were kept constant. Most of the species in the food web had a generation time of only several days, allowing hundreds to thousands of generations per species to be observed during the study period. Large fluctuations (over several orders of magnitude) in species abundance were witnessed, despite the constancy of external conditions. Different periodicities in species fluctuations could be discerned, such as a 30-day periodicity in phytoplankton-zooplankton oscillations, and the prediction of changes in short-term (several days) species populations was possible. However, species abundances also exhibited behaviour that was characteristic of chaos. In other words, the trajectories charting changes in species abundance fluctuated unpredictably. The presence of chaos in non-linear systems can be inferred by the use of Lyapunov exponents. Lyapunov exponents are explained in more detail later in the chapter, and so are only briefly considered here. Lyapunov exponents reveal the rate of convergence or divergence of trajectories over time, in this case the time-series trajectories of species abundance and resource variability. A system without a positive Lyapunov exponent means a convergence of nearby trajectories and indicates that the system is tending towards 'equilibrium' (e.g. periodic cycles in plankton species populations). In contrast, positive Lyapunov exponents, as exhibited in this food web, indicate divergence of nearby trajectories, which may suggest chaotic behaviour. Typically, this would result in an inability to predict longer-term changes in populations. Indeed, in their plankton food web, the predictability of plankton species populations was limited to 15–30 days (equivalent to 5–15 plankton generations depending on the species considered).

The implications of an inability to predict longer-term plankton populations could be profound. Beninca *et al.* (2008) highlight that many other food webs share this structure (plants, herbivores, carnivores and a microbial loop). In light of this, they tentatively suggest that a loss in predictability of species abundances in 5–15 generations could be a pervasive feature of many ecosystems. This research also revealed that despite the considerable instability in species populations, with large fluctuations in species abundances being observed over hundreds of generations, the food web persisted. Thus stability is not required for the persistence of complex food webs (Beninca *et al.*, 2008).

Chaotic behaviour in population numbers has important implications, both in terms of adjusting our view about how populations interact and change, but also for their effective management and conservation. Oscillations and chaotic fluctuations in species abundances may explain the so-called paradox of the plankton, a term coined by Hutchinson (1961) in reference to the co-existence of dozens of phytoplankton species in the face of a limited number of resources; a contradiction of competition theory, which maintains that the number of co-existing species cannot exceed the number of limiting resources (Hardin, 1960). The traditional view in community ecology is that complete competitors cannot co-exist; this is known as the competitive exclusion principle (Hardin, 1960). Complete competitors are two or more species from non-interbreeding populations that have the exact same resource requirements and share the same habitat. Huisman and Weissing (1999) investigated this issue further using a resource competition model in a constant environment, in which changes in phytoplankton species abundance were modelled as more resources were included. Competition over one or two resources produced predictable results – the system approaches a stable 'equilibrium'. If all the species are limited by the same resource, then the strongest competitor will displace the other species. Alternatively, co-existence of two species will occur if they are not limited by the same resource. Fluctuations in species abundance ranging from small cycles to large oscillations occur when three species and three resources are added to the model. A fourth species was added to the model – now the number of species was greater than the number of resources. However, the fourth species was able to co-exist with the other three species. Indeed, it was found that up to nine species could co-exist on only three resources. Co-existence was made possible because of the oscillations in species abundance, which were generated by the competition between three or more species for three resources. These oscillations may create an opportunity for additional species to exist.

When five resources were considered, the variations in population trajectories exhibited behaviour that was characteristic of chaos. Species fluctuations were irregular and the pattern of species replacement never repeated itself. Species abundance continuously diverged as several different species invaded at different rates when one species tried to become dominant. The model also demonstrated that species were sensitive to initial conditions – the trajectories of species with almost identical abundances diverged and became completely uncorrelated. The model also revealed that chaotic fluctuations in species abundance allow the co-existence of many more species than there are limiting resources. This is an extremely significant finding by Huisman and Weissing because it suggests that sufficiently complex plankton communities can generate non-equilibrium population dynamics, either oscillatory or chaotic, and, in this context, sustain multiple (i.e. co-existing) phytoplankton species on limited resources. Therefore, Huisman and Weissing argue, the solution to the paradox of the plankton may lie in complex changes in species abundances caused by competition. Moreover, the suggestion that, through competition, a more diverse phytoplankton community can be sustained, may serve as a suitable conceptual model for the biodiversity of many other ecosystems (Huisman and Weissing, 1999). In this sense, when oscillations and chaos are generated, competition is not a destructive force, but one that may allow a high diversity of species to exist on a limited number of resources (Huisman and Weissing, 1999). If chaotic dynamics are a potentially pervasive feature of biological populations then, as Costantino et al. (1997) caution, there are significant implications for the management and control of natural populations, where human intervention may trigger unpredictable and perhaps undesirable results.

Just stating that a system is chaotic and so unpredictable does not sound much like an explanation. It could be this negation of explanation, as some may interpret it, which is part of the resistance to accepting chaos and complexity theories. However, it could be argued that equilibrium is little better. Assuming a tendency towards a 'normal' state, even when dressed up with terms such as negative feedback loops, is still a constraint on explanation. Acceptance of inherent uncertainty as a valid form of explanation does seem to be almost anti-explanation. Likewise, the acceptance of chaos and complexity implies a change in what is defined as explanation in physical geography. Identifying chaos implies inherent instability in the operation of physical reality. It implies that a simple linking of one event to another, of linking a cause to an effect, may no longer be possible. Within systems analysis and the concept of equilibrium, a causal chain could be set up between events and outputs. The chain may be long or short, resistant to disruptions or sensitive to change. Within chaos, events now have multiple effects – singular patterns of cause and effect are potentially lost, and the context of the causes becomes much more important. A system close to chaos will produce different outcomes to one far from chaos. Although an argument could be made that equilibrium explanations have a similar type of change when the system is near a threshold, chaos regards these multiple relationships as inherent within the system; they are normal, they are not reflections of behaviour at the extreme end of the spectrum of system behaviour. In other words, chaos and complexity begin to redefine what is normal for a system. Identifying variables, their states and relations no longer guarantees that certain modes of behaviour will occur or can be predicted.

As well as unpredictable behaviour, concepts from complexity theory have also been applied to try to understand how order, in particular self-organisation, has developed (e.g. Dunne *et al.*, 2002 and Williams *et al.*, 2002 in relation to food webs). Early work by Kauffman on self-organisation of molecular systems indicated that at a critical threshold connecting entities within a system resulted in the emergence of 'organisation'. The example Kauffman (2000) uses is illustrated in Figure 8.3. Initially, the system of buttons or nodes is relatively weakly connected. After adding more threads or links a point is reached, as the ratio of threads or links to buttons or nodes reaches 0.5, where suddenly the whole network of buttons becomes interconnected. The sudden jump in connectivity was reflected, Kauffman believed, in molecular systems that were autocatalytic, that had self-reinforcing feedback loops for individual molecules. These molecular systems became very stable and able to resist changes in the environment and even act as relatively autonomous agents in the physical environment. Maintaining this organisation or structure, despite the perturbations thrown at the molecular system by the environment, illustrated how robust such self-organised networks could be. Rethinking other networks of entities in the same light, questions could be asked about their ability to maintain their structure when disrupted, and the relative importance of different types of nodes with different degrees of connectivity within such networks.

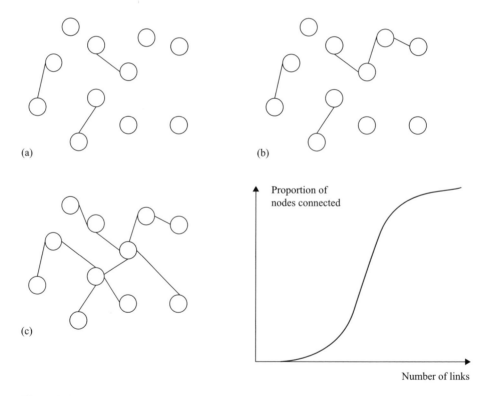

Figure 8.3 Initially a random network of buttons or nodes is linked by a few randomly placed links (a). As the number of links or threads increases, some small clusters occur (b). After a threshold value, a critical point (c), the proportion of the nodes or buttons connected increases dramatically; a change in the system has occurred, similar to that of a phase transition. Based on Kauffman (2000).

Emergence and hierarchies – scale revisited?

The development of chaos and complexity theories have, however, begun to highlight the importance of context for explanation. Additionally, the development of critical realism and pragmaticism has brought alternative explanatory frameworks to the attention of physical geographers. This has stimulated consideration of some old ideas within physical geography in a new light. Significantly, the role of context in explanation has surfaced as a key issue in contemporary physical geography.

The importance of reductionist forms of explanation cannot be underestimated within physical geography. However, like equilibrium, reductionism has undergone various changes in definition that make the establishment of its precise nature difficult to determine and the identification of a potentially alternative explanatory framework difficult to establish. Reductionism, as noted in Chapter 4, refers to the tendency to reduce explanation to the level of the lowest identified entities. Within physical geography this usually involves assuming that all 'real' explanation is located at the level of physics and chemistry. At this level, explanation is viewed as generalised and applicable to all entities. Increasingly, field sciences such as physical geography have found such sentiments of universality difficult to sustain given the questions of interest about reality in their subject. The problem seems to lie, as noted in Chapter 3, with the fixation of physical geography to provide what are perceived to be 'scientifically valid' explanations using the lowest entities available. Explanation is reduced to what are perceived to be natural kinds, which are therefore viewed as real entities, ones that actually do exist in reality.

Increasingly, use of and familiarity with philosophical positions such as critical realism and pragmaticism have highlighted the errors of this reductionist view of reality and have generated, amongst other things, a redefinition of reductionism. Critical realism, in particular, has enabled a reconceptualisation of what is an appropriate explanation. Critical realism recognises that reality is differentiated and stratified, as noted in Chapter 4. Each stratum of reality is composed of entities that are in turn composed of, but not reducible to, entities found at lower strata. Similarly, the entities in a particular stratum operate according to regularities, 'laws', that are not necessarily derivable from laws at lower strata. In other words, each stratum may have unique entities that interact according to unique sets of laws as well as obeying, or rather interacting, according to laws found in lower strata.

Explanation can be divided into the vertical and the horizontal. Vertical refers to explanation that focuses in on explaining one level by reference to laws at another level. This form of explanation looks for the underlying mechanisms that explain the specific power of a specific phenomenon. For example, the reaction of acid rain with calcium carbonate can be explained by reference to valency theory. Valency theory in turn can be explained by reference to interaction of atomic and subatomic particles. At the lowest strata level, the interaction of these particles can be explained by quantum physics. Extending this form of analysis to its logical conclusion would be a classic reductionist approach. Everything would be reduced to explanation at the quantum level. Unfortunately, at this level all entities are reduced to quantum particles. Building up from the level of quantum mechanics to that of the entity of interest is problematic. Some practitioners prefer to locate explanation at the level of fundamental mechanics and chemical reactions. Stopping at this level, it is often argued, is more appropriate. This implies that researchers have in mind an adequate level of explanation for their studies. Adequacy and appropriateness are therefore user-defined and capable of alteration.

Horizontal explanation is concerned with linking entities at the scale of study. This contextualises explanation. This level of explanation highlights the significance of the juxtaposition of different mechanisms, different causal structures in determining the actualisation and the power of a phenomenon. The contingent sets of relationships that define why something happens in reality (rather than how) forms the focus of this level of analysis. In the case of acid rain and calcium carbonate, the question is not so much how do they react, but why do they (H^+ cations and $CaCO_3$) occur in the environment in that location and at that point in time to be able to react?

Hierarchical-based explanations follow a similar differentiated view of reality and of explanation. Hierarchical explanations, such as that provided by Haigh (1987) within geomorphology, provide guidance as to what levels are appropriate for explanation. The level of study is viewed as providing the entities and relations that require explanation. The level below the level of study is the level at which processes and mechanisms are located, whilst the level above the level of study provides the context. This three-level view of explanation does provide a useful starting framework, but clearly problems arise over exactly how each level is defined. What it does highlight is that each level acts to mediate the actions of the levels below it, whilst being constrained in its actions by the levels above it. The series of complex interrelationships between levels provides a much richer and more diverse set of explanations than a single and simple view of links between scales.

This form of explanatory structure has similarities in the ideas put forward in hierarchy theory, used predominantly in ecology (Platt, 1970; O'Neill et al., 1980; Allen and Starr, 1982; de Boer, 1992). Within this form of explanation, it is accepted that reality is divided into different, distinct layers, or strata, as recognised in critical realism. The levels in the hierarchy interact and can influence each other. The rate of processes in the lower levels is much greater than the rate of processes in the upper levels of the hierarchy. Processes of photosynthesis, for example, are much faster than those of plant growth. Although the processes and entities of the upper levels are constrained by the processes operating at lower levels, they are not solely determined by, nor predictable from, the lower-level processes. Interaction, the development of 'laws' or rules of interaction specific to a particular level, also influence behaviour. Likewise, the upper levels can influence the lower levels by affecting the conditions within which the faster processes operate. Salt weathering may be rapid on a building, but it will not occur if the conditions are not appropriate for the combination of salt and moisture. Significantly, the researcher defines what the appropriate level of study is. Once this has been determined, it is then only the lower and upper levels immediately adjacent to the level of study that should be used in any explanation. Moving beyond these levels is reducing or generalising the explanation beyond what is deemed appropriate. Too general or too reductionist in mode and the explanation becomes unrelated to the specifics at the level of study.

Recent debate has focused on the possible existence of emergence as a possible means of explanation in physical geography (Harrison, 2001; Lane, 2001). Emergence, like most of the terms involved in explanation, is difficult to accurately pin down. Some researchers view it as referring to the tendency for new and novel entities or structures to emerge from the interaction of processes. These new entities are dependent upon the operation of processes and other entities for their existence, but their behaviour is not merely an aggregate of these other entities. Instead, the emerging entity has a coherence of form and function that define it as a distinct and separate individual at a particular scale. The emerging form can interact with other entities and respond as a single whole. The behaviour of the entity is greater than the sum of its parts. This view of emergence sees entities as forming at distinct scales as a response to processes at other levels. Action at a particular scale is not dependent upon processes at a smaller scale, but on the emergence and behaviour of emergent entities. Other researchers view emergent entities as being distinct individuals at a particular scale but that these individuals are explainable by reference to the processes that formed them. The entities may act as individuals, and even interact with other individuals at that scale, but their behaviour is an aggregation of processes and entities found at a lower scale. In this view of emergence, entities emerge, but their behaviour and existence is predictable from our knowledge of their aggregate components. Such distinctions are vital because in the latter view of emergence, explanation is placed at the level of the components, whilst in the other view of emergence, where entities are irreducible, explanation is placed at the level at which the entity emerges.

The above views of entities and emergence have echoes in the writings of A.N. Whitehead (e.g. Whitehead, 1929). Whitehead viewed entities as being temporary concretisation of flows of processes. Processes became solidified in the act of creating entities. The entity only remains, however, if the flows themselves remain constant. The entity could affect the flows and so enhance the solidification of its form. Any single entity though was connected to all other entities by virtue of the flows that formed it. This meant that a single entity was connected horizontally to other entities at a scale similar to itself. Likewise, entities were formed by processes that were governed both by internal mechanisms and by their context.

Combined, this view of emergence begins to ask interesting questions about scale and its definition in physical geography. A common assumption is that scale is concerned with an absolute set of spatial and temporal dimensions. Entities have an existence relative to this fixed frame of reference. Different entities exist at different magnitudes within this fixed frame of reference (Figure 8.4). Research by Raper and Livingstone (1995) in relation to geographical information systems, and by Harvey (1994) and Massey (1999) in relation to geography in general, has begun to suggest that scale may not be as clearly defined or fixed as previously thought. As with Whitehead's philosophy, a more relational, or even relativistic view of scale, may be more appropriate to physical geography. Rather than scale being absolute within a fixed reference frame with the entities fitting within this, scale could be thought of as being defined by the entities themselves. Scale is no longer absolute but is dependent upon the entities under study. The entities themselves define the processes or flows forming them, they define the spatial and temporal dimensions of importance rather than being defined by these. Scale is no longer an absolute quantity, but one that varies with the entities being studied and so is defined by the study itself. This is rather different from the static spatial and

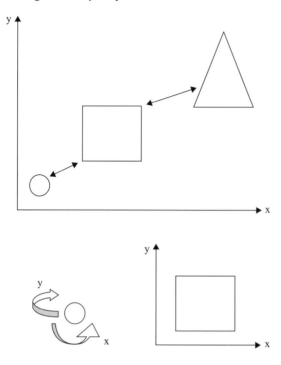

Figure 8.4 Entities are placed within a fixed x, y co-ordinate system. If an entity develops over time from a circle to a square there is no change in the reference frame used. If the circle entity defines the nature of the x, y co-ordinate system, then development into a square with its own associated co-ordinate system implies a change in the nature of the co-ordinate system within which description and analysis takes place.

temporal frameworks provided by Schumm and Lichty (1965). Within their framework, space and time are absolute quantities within which variables vary in significance. The fixed spatial and temporal frame provides a common and unfaltering backdrop for the action of variables. By turning the whole structure on its head, the spatial and temporal framework become relative, defined by the variables themselves. This provides a more dynamic view of the framework within which variables and relations operate. The scale of operation of a variable is defined by its relation to other variables, not by a fixed framework.

Phillips (1988, 2001) also looks at the issue of scale within physical geography. Following the work of Schaffer (1981), Phillips (1986, 1988) notes that it can be demonstrated mathematically that in systems where interactions operate over timescales of greater than an order of magnitude, these can be thought of as being independent.

> Controls acting at any given scale can be considered to be an abstracted subset of all system components operating at all scales. If the abstracted and the omitted relationships operate over spatial scales an order of magnitude different, the relationships are independent of each other in terms of their influence on system behaviour.
>
> (Phillips, 1988, p. 316)

This is a similar argument to that put forward for a hierarchical view of reality above. In this view, processes at the lower levels or strata operate at rates so fast as to have little influence on the upper strata. Likewise, the upper strata, although constrained by the lower strata, operate at their own rates and have little impact upon the state of the lower strata. Each level has its appropriate conditions for stability that may only be influenced by the other strata in exceptional circumstances. One of these circumstances could be the presence of chaos or complexity within and between the levels.

Importantly in Phillips (2001), this recognition of scale differences is combined with a recognition that this means that there is unlikely to be a single, best representation of reality across all scales. An appropriate representation at one scale, derived from a specific method, will not form an appropriate representation at another scale. A micro-scale form may most appropriately be identified and measured by a particular method-ology. This same methodology will be unable to represent the phenomena that the entity represents across all scales. As the scale of study changes, the nature of the phenomena changes and the means to represent it in an appropriate manner alters. The problem becomes, as Phillips notes, one of trying to reconcile unsuccessfully different and some-times competing methodological approaches. This is not a problem if it is accepted that reality is not capable of continuous and singular representation.

> Any frustration and despair arising from accepting the argument that there are fundamental limits on the ability to use single representations across scales would likely come from reductionists confronted with the realization that their methods are not always sufficient, or even worthwhile, for addressing some problems. Proponents of historical and system-oriented approaches have acknowledged that their methods are often insufficient and sometimes useless, and have generally recognized and accepted that historical or systems approaches are best suited to particular ranges of time and space and ill-suited to others.
>
> (Phillips, 2001, p. 757)

The scale changes, and so does the means of representing phenomena. The entities that reflect the phenomena have to alter as the means by which they are identified and measured alters. Entities and scale co-vary. It is pointless to view scale as independent from the entities being measured, the two cannot exist independently.

Case Study

Scale and (dis)connectivity

Fryirs (2012) looks at (dis)connectivity in catchment sediment cascades in the context of the sediment delivery problem, first identified by Walling (1983). Put simply, the problem is that only a fraction of the sediment eroded within the catchment makes it to the catchment mouth or outlet, or at least to the point at which the measurement of sediment is made. The rest of the sediment is retained within the catchment and research has focused upon how different storage mechanisms are activated (or not)

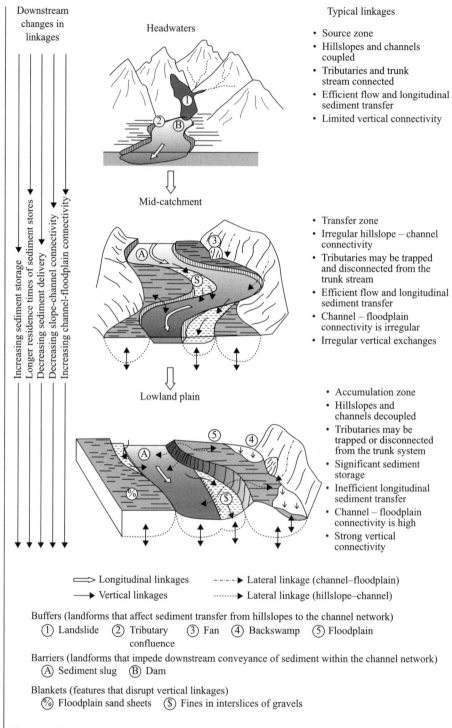

Buffers (landforms that affect sediment transfer from hillslopes to the channel network)
  ① Landslide ② Tributary ③ Fan ④ Backswamp ⑤ Floodplain
           confluence

Barriers (landforms that impede downstream conveyance of sediment within the channel network)
  Ⓐ Sediment slug Ⓑ Dam

Blankets (features that disrupt vertical linkages)
  %̸ Floodplain sand sheets Ⓢ Fines in interslices of gravels

Figure 8.5 Linkages in an idealised catchment. Linkage nature and strength varies with location. From Fryirs (2012).

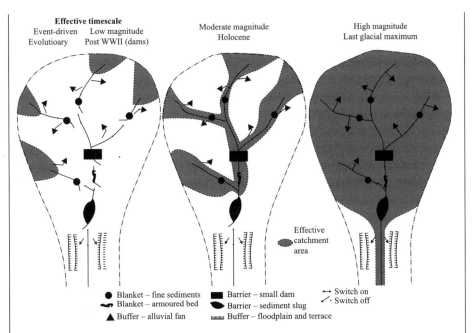

Figure 8.6 Switches in a catchment. Effective catchment area is the spatial area contributing to or transporting sediment along the fluvial network. Effective timescale refers to the magnitude of the event required to overcome buffers, barriers or blankets and reflects the timescale over which disconnectivity occurs. The figure illustrates that the processes, and in particular events, define the spatial and temporal framework of reference and the fluid and relational nature of this framework. From Fryirs (2012).

to retain this sediment. Fryirs presents a conceptual framework that uses the (dis) connectivity in space and time of different parts of the catchment to understand the variability of sediment delivery. She identifies buffers, barriers and blankets as specific types of storage and linkages based on landscape configuration, which can affect response times of the catchment.

The store types and linkages are identified and summarised in Figure 8.5. The operation of the conceptual framework outlined by Fryirs could be reinterpreted in terms of different space-time frameworks of landform. The differences in storage types and the density of linkages between storage types vary with the position of a landform in the catchment. This means that different parts of the catchment will be responding at different rates and so exhibit a different space-time context for their analysis. In addition, this space-time context will impact on other landforms further down the catchment as the release or storage of sediment will determine the space-time framework of operation downstream.

Fryirs identifies the difference in timescales when she analyses switches in the catchment as in Figure 8.6. The effective catchment area is defined as the spatial area within the catchment that contributes to, or transports, sediment along the fluvial system. The effective timescale is defined as the magnitude of the event needed to overcome the storage areas and types. The storage areas and types effectively produce a disconnection between different parts of the catchment, which the 'switching' overcomes. Breaching the storage areas reconnects different parts of

Holocene – highly disconnected

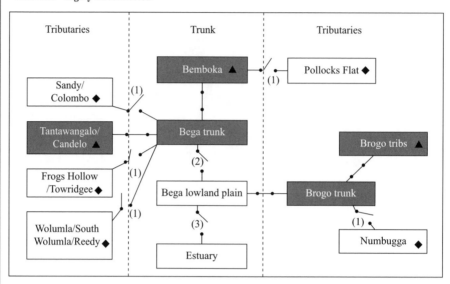

Post European settlement – highly connected

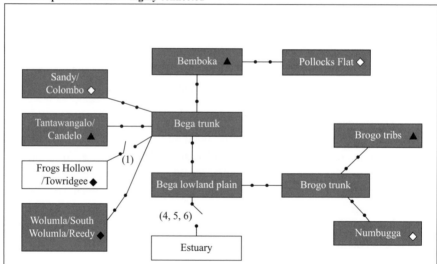

Buffers, barriers and blankets
(1) Intact valley fill and floodouts
(2) Bedrock step
(3) Continuous floodplain
(4) Sand slug
(5) Floodplain sand sheet
(6) High width:depth channel

▬ Effective catchment area

◆ Tributaries with swamps
◇ Tributaries with incised channels
▲ Tributaries with bedrock-controlled channels

Figure 8.7 Illustration of nature of disconnectivity in Bega catchment pre- and post-European settlement. The figure is represented as a 'switch' diagram where tributaries with swamps, incised channels or bedrock-controlled channels contribute materials to the main channel and the estuary. The 'on-off' nature of the switch, and so the level of disconnectivity, has altered due to human disturbance associated with European settlement. From Fryirs (2012).

the catchment in terms of the movement of sediment. This means that the spatial area of the catchment acting or responding as a unit changes with the magnitude (and potentially frequency) of the triggering or breaching events. This could be reinterpreted as event defining the scale of the entity being studied, with the entity being at its greatest spatial extent when a high-magnitude, low-frequency event occurs as in Figure 8.6.

The importance of (dis)connectivity for defining the spatial and temporal scale of the entity under study is also illustrated in Figure 8.7, which shows the nature of connectivity in the Bega catchment before and after European settlement. Fryirs views the example as illustrating the relative nature of space and timescales in entity definition, although it could also be argued that the figure represents the relational definition of entities. The impact of an event is dependent upon the timescale of reference. At relatively short timescales the event magnitude and frequency, the sensitivity of landforms and their recovery characteristics all affect how the different parts of the system became connected (or not) and so affect the nature and scale of the landform or landforms that are connected. As the magnitude of the event increases, then the interlinkages between different parts of the catchment are more likely to be activated and so the spatial and temporal scale of the entity affecting sediment delivery increases as well. The nature of the linkages, whether they enhance or dampen sediment movement, will determine the nature of the change in sediment delivery, but the important aspect is that the scale of the entity is not predefined by absolute spatial and temporal dimensions. Scale is, instead, defined by the varying relationships activated between different entities and parts of the catchment. In this sense scale is relationally defined.

Case Study

Complexity and change – landscape evolution and organisation

Phillips (1995) outlines how the concept of non-linear dynamical systems (NDS) can be applied to the analysis of relief evolution. The key question he is trying to answer is: does landscape evolution exhibit deterministic chaos? (Phillips, 1995, p. 57). Deterministic chaos would result in complex, irregular patterns resulting from deterministic systems regardless of the presence of other factors such as stochastic external forcing factors or confounding factors. Phillips considers the non-linearity in landscape development to arise from a non-proportional relationship between inputs and outputs resulting from changes in stores and system thresholds.

Phillips suggests that deterministic chaos would leave a signature magnifying the imprint of disturbances of any magnitude and so produce complex (or convoluted) topography. Without deterministic chaos small disturbances would be irrelevant and quickly removed as the topography converged on a steady-state standard topography. Stated simply, this means that if relief increases over time then chaotic behaviour is implied, whilst if relief decreases over time then non-chaotic behaviour is implied. It is important to recognise that Phillips does not state that relief divergence (increasing difference over time) is indicative of chaotic behaviour, only that it could be. He recognises that other factors could cause such behaviour, including stochastic forcing factors.

In a NDS view of the landscape there will be n interacting components, $x_i = 1, 2,$ . . . . n. Over time the behaviour of any of these individual components could be expressed as a function of other components (x) or parameters (c) in the form of an ordinary differential equation:

$$dx_1/dt = f_1 (x_1, x_2, \ldots x_n, c_1, c_2, \ldots c_m)$$

In an ideal world, everything would be known about the system, every component and every parameter would be known and would be expressible as above. If this were the case, system change could be described by mapping these equations in n-dimensional phase space. In reality, such complete knowledge is not possible, so a smaller subset of the n-dimensional phase space is plotted: q-dimensional phase space. Phillips introduces the concept of phase space to help in his description of system behaviour, but it is also worth noting that the recognition that a smaller subset is used in reality to map system behaviour raises questions about component and parameter selection. How representative is the subset of the larger phase space? And if other components and parameters were selected, would the description produced change?

Whether the landscape has a tendency towards convergence or divergence of relief can be determined by looking at the Lyapunov exponents (λ). Lyapunov exponents of a system are a set of invariant geometric measures that indicate system dynamics. If you visualise system behaviour as a set of trajectories in phase space, then the Lyapunov exponents quantify whether trajectories converge or diverge, and how rapidly they do this (Figure 8.8). Where the average Lyapunov exponent is positive, then the trajectories diverge, when it is negative they converge, when it is zero there is no divergence or convergence. The average Lyapunov exponent refers to the general behaviour of the system. A positive value only implies that chaotic behaviour will be exhibited, not necessarily that chaotic behaviour will always be exhibited. Where positive Lyapunov exponent values occur, this implies a potential for chaotic behaviour with the largest positive value indicating the potential rate of divergence. Calculation of Lyaponuv exponents is relatively straightforward if all the differential equations describing the system components and parameters are known. In reality, this is never the case. Instead, Phillips makes use of a useful property of NDS, that the

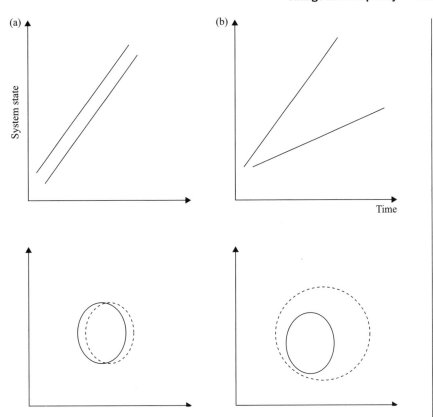

Figure 8.8 Visualising Lyapunov exponents. Graph (a) shows the shape and location of the components remaining almost the same for the entire trajectory. (b) is a pair of diverging trajectories. The components show an increase and expansion of the shape represented in the graphical phase space. Axes on the graphs are two-dimensional representations of the components.

dynamics of the system can be derived from a single observation. This observation integrates the mode of system behaviour and charting its change over time will reflect the behaviour of the NDS as a whole. If the NDS is chaotic then randomly chosen pairs of initial conditions will diverge exponentially at a rate equal to the largest Lyapunov exponent. Mathematically, once two trajectories become separated it is increasingly difficult to describe the trajectories using the Lyapunov exponents, so every so often the variations in trajectories need to be rescaled, or better still, renormalised to produce trajectories that are close together.

For landscape development the theoretical terms can be translated into components capable of assessment. Phillips uses two theoretical points in the landscape, i and j, with elevations h_i and h_j at two times, t_o and t. The elevation difference is given by h_i-h_j and the rate of change is:

$$d_{i,j}(t_o + t) = [h_i(t_o + t) - h_j(t_o + t)] = [(h_i(t_o) + h_i) - (h_j(t) + h_j)]$$

For this representation of the landscape and its dynamics, a positive Lyapunov exponent is present when:

$$d_{i,j}(t_o + t) - d_{i,j}(t) > 0$$

In other words, if the rate of change at time t is greater than the rate of change at time t_o then the system has a positive Lyapunov exponent and is chaotic.

Using this relatively simple means of identifying chaotic and non-chaotic behaviour, Phillips defines ten types of behaviour for relief (see Table 8.2 and Figure 8.9): five stable and five unstable. Phillips notes that the NDS model of topographic evolution could provide a unifying framework for all existing theories about topographic evolution. All these theories could be mapped onto the ten modes of change making the NDS model itself unfalsifiable.

Table 8.2 *Relief behaviour, from Phillips (1995).*

Behaviour	Description
Stable 1	Planar surface experiencing spatially uniform rates of erosion, accretion and uplift
Stable 2	Both points eroding, with erosion rates at the initially higher point greater than or equal to that at the lower point
Stable 3	Both points accreting or being uplifted, with rate of increase in height at initially higher point less than or equal to that of the lower point
Stable 4	Initially higher point is eroding and lower point is accreting or being uplifted
Stable 5	Higher point is not changing and lower point is accreting or being uplifted
Unstable 1	Planar surface with any variation in erosion, deposition or uplift rates
Unstable 2	Both points are eroding, with rate at the initially higher point less than that at the lower point
Unstable 3	Both sites accreting or being uplifted, with rate at the higher point greater than at the lower point
Unstable 4	Initially higher point is being uplifted or accreting and lower point is eroding
Unstable 5	Higher point is not changing and lower point is eroding, *or* lower point is not changing and higher point is accreting or being uplifted

Despite the unPopperian status of the NDS theory, Phillips does think it provides some basic tenets for a model of landscape evolution. These are, first, that there are ten modes of topographic evolution differentiated by the average rates of uplift/accretion or erosion/subsidence at initially higher and lower points. Second, that the five stable modes involve constant or declining relief over time, whilst the five unstable modes involve increasing relief over time. Third, neither stable nor unstable modes can persist over geologic timescales or across the whole landscape. Fourth, chaotic modes involve sensitivity to initial conditions and inherent unpredictability, but within well-defined boundaries associated with relative rates of change in the landscape. Finally, planar surfaces are unstable.

Phillips's view of NDS as a potential integrative theory for landscape development is an interesting application of ideas from chaos theory. The development of a

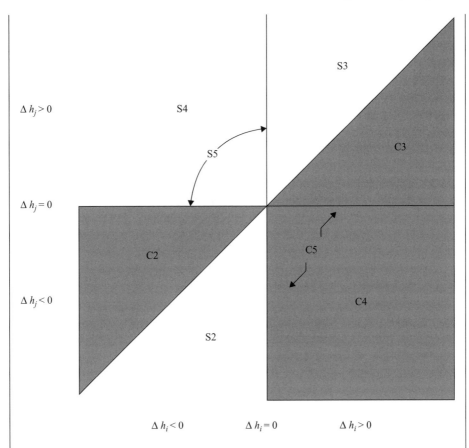

Figure 8.9 Behaviour of points in different relief models, based on Phillips (1995). Stable (S) and chaotic (C) modes of topographic evolution in the NDS conceptual model. S1 and C1 apply to initially planar surfaces and are not shown. The subscript i represents initially higher and j initially lower points in the landscape. S5 and C5 plot along the principal axes as shown.

mathematical basis to his argument permits him to identify some indicators of chaos within systems, provided that the theoretical concepts can be appropriately translated into real entities capable of measurement. It is problematic that the theory predicts almost every mode of landscape development as this makes the falsification of the theory virtually impossible. Any pattern of behaviour is permitted by the theory and so every pattern of behaviour can be explained away. NDS does, however, provide a relatively easy to understand interpretation of landscape development using almost the simplest of systems: two points. The range of behaviour that these two points exhibit can be interpreted in terms of stability and change; in this light, does the framework of NDS provide any more explanatory power than the traditional framework of equilibrium?

Summary

Change and stability in the physical environment have been explained via systems thinking. Equilibrium has been used as a reference point for different types of behaviour, even alternative states to equilibrium are defined relative to it. Equilibrium is a term borrowed from other sciences and applied in a manner that makes it very difficult to disprove. The concept can be translated in a range of ways depending on the circumstances. The use of equilibrium as an explanation has resulted in the expectation of certain types of behaviour and constrained research to look for these forms of behaviour. The development of complexity as an alternative to equilibrium has suffered from many of the same problems. Depending on the interpretation, complexity and chaos theory can be interpreted as including equilibrium as a form of expected behaviour. Terms such as self-organisation are capable of translation to reality in a number of ways and so are difficult to test as explanations. Similarly, identification of chaos or complexity relies on the same measurement systems used to identify equilibrium. Separating the two concepts and devising decisive tests has been problematic.

Chapter 9

Modelling

Models refer to the conceptual or mathematical simplification of reality; they are constructed in order to understand some aspect or process. For some aspects of physical geography such as climatology and understanding the relationships involved in chaotic and complex systems (Chapter 8), modelling is an essential tool to understanding; it could be argued that modelling is the *only* means available to facilitate our understanding of reality in these areas. Models are not reality itself and should not be viewed as such. Researchers who use models do not necessarily believe that their models correspond directly to reality. Nevertheless, on occasion some researchers may place too much confidence in models, and, likewise, some decision-makers may behave as if the models are reality. If this were not so then the shock associated with the failure of a model would not be such a surprise. Harvey (2010), in relation to the ongoing economic crisis, suggests that economic theorists (citing Samuelson in the *Washington Post* as his source) 'became too interested in sophisticated forms of mathematical model-building to bother with the messiness of history and that this messiness had caught them out' (p. 235). Add 'space' and 'place' to history and you have the nub of the problem. Models are simplifications, potentially complicated simplifications, but still simplifications of the messiness of reality. Belief in models as reality is a very correspondence-based view of reality (Chapter 2) and by implication the practice of science.

Leaving this reservation aside, the central role and importance of modelling in contemporary physical geography is paramount. Without modelling, whole areas of the subject would not exist or would at least have difficulty in developing understanding of their subject area. Demeritt and Wainwright (2005) suggest that models provide a means of understanding and predicting the operation of systems that cannot be analysed through experimental methods such as variable control and manipulation. In addition, analysis may not be feasible for practical and political reasons or because of the spatial and temporal scales of the systems involved, and so modelling techniques have to be used instead. General circulation models in climatology are a clear example of this. Acceptance that modelling represents reality (even if researchers know and understand the limitations of such a view) has also been important for the development of the subject. The development of an accepted standard model of climatic change in the UK over the next 30 years (UKPC09) has been the basis of research projects and the development and refinement of our understanding of the likely impacts of future climate change. It is also probably correct to say that without modelling, or the acceptance of modelled behaviour as real as in the case of UKPC09, funding would be almost impossible to come by for some areas of the subject.

This chapter will discuss modelling by examining conceptual approaches to modelling, the types of models physical geographers use and then explore some of the key issues that confront those trying to model reality or understand reality through their models.

Conceptual approaches to modelling

Levins (1966) viewed modelling as an attempt to maximise generality, realism and precision of models in representing the real world so as to achieve the aims of understanding, predicting and modifying reality. Levins views these aims as overlapping and not capable of being achieved simultaneously. Constructing any model required a trade-off between these aims, thus the developing of an accurate model of reality in every respect is not seen as a possible goal by Levins.

There have been discussions within modelling as to the philosophy and nature of the subject. These discussions have tended to view modelling as a means of explaining and predicting the physical environment. They tend to be based on the Hempelian view (Hempel, 1965) that explanation and prediction are logically identical. A model that can explain reality can also predict it and so by extension if a model can predict reality it must also explain it. Explanation then becomes a case of understanding the underlying structure and relations outlined in the model. The model, rather than reality itself, becomes the focus for explanation, but reality is seen as being explained by the model. A correspondence view of reality is implicit in this approach; the models reflect directly what reality is like even if in some cases there is an acceptance that technology is not quite up to modelling reality as well or as accurately as we would like. An extension of this view is that the entities and relationships that make up the model are themselves 'real'; they have existence beyond the modelling process itself.

Beven (2002), for example, although stating that models can be produced without any explicit underlying philosophy, outlines his view of environmental modelling as a form of pragmatic realism. He suggests that most practitioners have a view that, despite the constraints of current knowledge, computing power and observations, the models they develop are as close to reality as they can be. Refining and developing these models to improve this closeness has generated vast research programmes and vast amounts of funding. This view is, he suggests, a pragmatic realist view of modelling which, although naïve, is one that aids in the task of developing models that mimic reality but which are known to be wrong due to the constraints mentioned above. Beven suggests that the modeller has a clear perceptual model of reality that provides qualitative understanding gained from experience, training and monitoring although this model itself is limited by current constraints. Applying this perceptual model requires practical application of this understanding through the development of a conceptual or formal model.

Beven (2002) cites Cartwright's views (1999) on nomological systems as a conceptual basis for modelling. Cartwright argues that representation of the capacities of real entities can only be made through nomological systems with their own defined constraints. Beven argues that building a nomological system may require accepting that that system will not be consistent with the perceptual model the modeller has of reality but such a system will be required to produce predictions from the model. It seems that Beven views the nomological systems as an approximation to reality that is related to the perceptual system the modeller has of reality, with the implication that the latter is somehow more realistic than the former. There seems to be a mix of correspondence and coherence views of reality within this one conceptual approach.

Beven (2002) also explores the ideas of landscape space and model space as a conceptual framework for modelling. The 'landscape space' represents reality and cannot be fully known as, Beven states, our perceptual models are currently inadequate and change though time. 'Model space' can be known perfectly as it represents the space defined by our deterministic models with known boundary conditions and mathematically modelled relationships. Landscape space may throw up events that alter our understanding and so impact on the nature of the models and therefore extend the scope of model space. Modelling becomes the process of trying to match, as accurately and precisely as possible, the model space to the landscape space. The process of modelling may reduce a complex aspect of landscape space to a single output in model space but this at least provides a degree of understanding of landscape space in terms the modeller can relate to and translate to the terms of their model creation. Further developments in technology and modelling will eventually enhance this understanding and the full nature of landscape space will unfold as the model space evolves and improves.

Within the above view of modelling there is potential for model rejection based on its match to landscape space, for model refinement and for model improvement as the constraints on modelling change. Conceptually, there are a number of points that need to be considered. Cartwright's nomological system recognises the capacities of entities but the clarification of their reality or ontological status is not clear-cut. Mapping model space onto landscape space is an interesting metaphor but it is unclear how the researcher knows that they are mapping appropriately. Landscape space is unknowable as it represents reality; model space is a human construct and should be known perfectly (although this could be questioned when non-linearities arise or if the carefully defined boundary conditions are violated). Translating from model to landscape space seems to carry the implication that the modeller has managed to match reality. This further implies that the model therefore provides an exact description and even an exact understanding of reality. Given enough resources, technological development and time, model space will map exactly to landscape space. Whether this means we will have an exact description of reality or an exact understanding is unclear.

Modelling has, traditionally, been grounded in developing models from 'first principles', the classic reductionist view that models need to be based in classic physics and chemistry and that only these types of models provide an appropriate basis for understanding and predicting physical systems. Integrating into a single model as many of these 'fundamental principles' as possible has been viewed as the way to improve models as the more of these that can be incorporated into the model, the more 'complete' the model becomes. This view of modelling sees the fundamental principles, particularly the equations or expressions that describe these entities and their relationships, as the cornerstone of understanding. Reality, in this view, can only be described and understood via these fundamentals. These fundamentals represent reality accurately and so when you build models based on these you are building models on the basis of reality as it is. Murray *et al*. (2009) views such models as simulation models in that they are designed to reproduce natural systems as completely as possible.

More recently an alternative 'synthestist' school of modelling has developed in geomorphology (Paola, 2001). This approach to modelling is based on the view that the complex entities and complex relationships that make up reality do not necessarily require complex modelling based on integrating fundamental underlying principles. The synthestist approach recognises that complex systems operating across scales may be dominated by only a few crucial aspects at different scales (Paola, 2001). Synthestists have a critical realist view of reality in that they recognised that reality is hierarchical and that lower levels in a hierarchy provide the basis or rules for behaviour for a higher level in the hierarchy. Modelling, therefore, becomes a process of identifying the level of the hierarchy that is the focus of research and then determining the relationships or dynamics at the lower level that are important for the operation of the level of the hierarchy of interest. These 'rules' from these lower levels in the hierarchy may represent simplifications of some 'fundamental principles' or they may be constructed as completely new artefacts at that level. Such models are only 'reductionist' in the sense that explanation (and prediction) is reduced to the lowest level that is appropriate to explain (or predict) the entities and their relationships at the level of research. The 'simplified' relations at this lower level are researcher-defined, even if they are based on 'fundamentals', and so are explicitly the product of the training, experience and conceptual frameworks of the researcher. Modelling becomes a process of understanding reality, as far as we can know it, through entities and relationships at a level appropriate to, and couched in terms comprehensible to, the needs and knowledge of the researcher.

Types of models

Modelling in physical geography can be traced back beyond the so-called 'quantitative revolution' of the 1960s and 1970s, with the likes of Strahler (1952, 1957) exploring mathematical methods in physical geography as well as an ongoing debate about the 'scientific' credentials of geography expressed through Schaefer (1953) and Hartshorne (1959).

Substantive tomes, such as Chorley and Haggett's *Models in Geography* (1967), provide the basis for claiming the 'quantitative revolution' as the foundation for modelling in physical geography. The impact of this period and these works on the development of research and on the thinking of a generation, if not generations, of physical geography students cannot be neglected. Equally, the opposition to modelling found its voice during the same period of growth, so both modelling and its criticism developed hand-in-hand. The drive to model has many underlying purposes but a key one for the early modellers was the belief that modelling offered a way to unify physical and human geography under one common philosophy and methodology.

Chorley and Haggett (1967) provide a set of definitions for what a 'model' is that still provide a useful starting point for any discussion. Simply put, a 'model' is a simplification of reality, a researcher-defined representation or abstraction of what they believe to be the real world. Giere (1988) defines models as idealised structures that researchers use to represent the world though resemblances between the model and the reality. These broad definitions encompass a wide range of types of models, some of which might not even be considered to be models or represent modelling. This definition does, however, highlight the key issue discussed above – a model represents something, it is *not* the thing itself. Any discussions of the model, how it changes or remains the same, how it responds to changes in inputs, how sensitive it is to changing environmental and internal conditions, all these discussions refer to the model, not directly to the thing it represents. Manipulation of the model and the resultant changes observed are changes expressed by the model; whether these changes are or will also be expressed by the thing the model represents is another issue. A good model will reflect the thing it represents well and it might be expected that the thing will behave in the same way as the model. For brevity of expression it may even be that the model stands for the thing it represents in discussions, but the model is a model and not reality.

There are, however, also different types of models or rather different ways that models can represent reality. These different types of models reflect both different ways to model reality as well as different views about the nature of reality. Chorley and Haggett (1967) also provide a typology of models summarised and discussed below.

Conceptual models

In White *et al.* (1992), when systems are constructed, variables selected, relationships identified and lines made to represent reality they are consistently referred to as 'system models'. The system, with all its component parts, is a simplification of reality designed by the researcher and it represents a conceptual model of reality. Conceptual models reflect the researcher's view of how reality, or rather the variables and relations they have identified as essential to the operation of the section of reality they are investigating, fit together. In this manner, the conceptual model reflects the researcher's underlying theory or theories about the operation of the physical environment. Typically, in conceptual models within physical geography this involves drawing a systems figure as illustrated in Chapter 7. The figure reflects how the researcher believes reality to be structured and the links between boxes, reflecting the relationships or processes the researcher believes need to be studied in order to understand the dynamics of the system. Conceptual models, therefore, drive the manner in which research is undertaken. They direct the researcher as to what variable to measure in the field, the conceptual basis for that measurement as well as how the measured variables should impact upon other elements in the system. These simplifications of reality determine how reality is identified and quantified and so their existence partially determines how reality is understood. No understanding of reality can be achieved without some model to guide the investigation of the researcher but by the same token the model itself underpins that investigation.

Model construction tends to be based on some underlying theory of the phenomenon under investigation. The model takes the abstract concepts and entities and tries to produce a set of variables and relations capable of identification and measurement in the field. The model provides the basis for bridging the gap between theory and practice. The translation may not be perfect as pointed out in Chapter 2 and the discussion of bridging principles, but by explicitly connecting theory to reality the model does provide a basis for further manipulation of the model to assess whether the phenomenon is present in reality and, if so, how its influence varies with changing conditions. The model provides the template for critical manipulation of the phenomenon.

Analogue models

Analogue models have a long tradition within physical geography (e.g. Chorley, 1964). Analogue models are physical systems that are viewed as behaving like, or have a form that is analogous to, the system under investigation. A researcher could build a hardware model of a fluvial system using a flume, sediment and flowing water and believe that this is an adequate physical analogue model if it captures what they believe to be the key elements of a fluvial system relevant to their research. Such a model implies not only that a researcher has a model of the system in mind but also that this system has variables that are so influential that their behaviour can be adequately modelled by a physical surrogate. Hardware models could be viewed as being a type of analogue model as well.

Issues of scale representation (spatial and temporal) arise when hardware models are constructed as other than a 1:1 scale model, there are bound to be spatial and temporal distortions formed in a hardware model. Despite these issues, the analogue models do provide a basis for illustrating and, to a limited extent, analysing the basic principles that underlie the physical system under investigation. Similarly, the use of analogue models highlights a key issue in scientific analysis – the representation of one system by another – the argument by analogue or metaphor (Hesse, 1966). Analogue models work because they 'stand-in' for the physical system of interest. The analogue is deemed to be sufficiently similar in the important respects to the researchers to be a valid representation of the physical system for analysis.

Deterministic models

Formalising models using mathematical expressions to represent relationships between elements is a key feature of current modelling practice. Deterministic models usually try to model systems from basic principles. Working from this basis, relationships are deduced and the operation of the resultant model can be explained by reference back to these basic principles. The important aspect of this type of modelling is that the relationships must be formalised as mathematical expressions. The behaviour of the resultant model is explainable by reference back to the basic principles and their formal relationships. For example, in ecology, Lotka-Volterra models of interacting populations use mathematical descriptions to model changes in population. Such models are limited, however, because they concentrate on modelling the interaction between organisms without consideration of feedbacks between the organism and their abiotic environment (Wilkinson, 2006).

Deterministic modelling relies upon any variable or entity and its relationships being expressed or reduced to a set of basic and fundamental physical principles. This means that it is essential that abstract axioms are linked to real-world entities by bridging principles as outlined in Chapter 2. Understanding the nature of this translation and its validity is vital for the acceptance of the model as an appropriate simplification of reality. This can become problematic as the scale and number of relationships defining the reality under study increases. Increasing the scale and complexity of the entities beyond the usual domain of the 'hard' sciences introduces the question of the use of appropriate 'laws' or relations for the scale of analysis. A critical realist view of reality, as outlined in Chapter 3, would view reality as being differentiated and stratified, and so the use of 'laws' derived from a 'simpler' level of reality to totally explain the behaviour of entities at a more complex level would be viewed as problematic. A solution could be to describe 'fundamental laws' of principles as being scale-dependent or scale-appropriate. This means that rather than reducing the variable of entity to its basic principles as defined by fundamental physics and chemistry, the variable or entity is described and modelled using the physical and chemical relationships appropriate to the scale at which it is identified. This seems to be the manner in which modelling is carried out and so an implicit critical realist view of reality seems to prevail.

Empirical-statistical models

Empirical-statistical models use statistical methods to obtain mathematical expressions that are meant to represent the physical system under study. Often the relationships modelled are ones of regression derived from existing data sets. In this way dependent variables are modelled by independent variables and causation is implicit within the model structure. The requirement of an underlying data set as the basis for this form of modelling means that data have to be available in a format appropriate to the model and its definition of variables and relationships. Data sets could, therefore, drive the types of models developed as well as providing the basis for validating those models as well. These types of models are sometimes dismissed as being simple 'input-output' models, matching input variables to output values through the development of a mathematical expression. Although there may be a grain of truth in this belief, the model is not usually constructed without some recourse to physical processes and what the modeller understands about them. This implies that there are physical processes linking rainfall to streamflow even if the whole complexity of these relationships and intermediate processes and states cannot be explicitly stated. Although these forms of models do not explicitly state why the input and output may be linked, the selection of the inputs does imply a set of processes and a view of the physical system and its operation. Likewise, even if the system and processes cannot be stated accurately, the fact that the model seems to produce predictable results may be sufficient for the model to function adequately in its particular context.

Case Study

Numerical modelling of Late Quaternary relative sea-level change and glacial isostatic adjustment

Now that the different types of models and approaches to modelling have been introduced, and before we consider the issues surrounding modelling in the following section, it would be useful at this point to review in detail the application of models, both to highlight their importance and the potential insights that they can provide, but also to reveal their potential problems. Our example focuses on the modelling of glacial isostatic adjustment, an area in which significant progress has been made in recent years.

Glacial isostatic adjustment (GIA) models attempt to capture spatial and temporal changes in vertical crustal motion and associated changes in relative sea level (RSL; defined as the height of the ocean surface relative to the solid Earth). The time period from the Last Glacial Maximum (LGM; *c.*21 ka BP) to the present day is often targeted for study by the GIA modelling community because of the significant redistribution in Earth's surface mass during this interval (e.g. changes in ice mass and water-mass loading). Since the LGM, sea level has risen by over *c.*120 m (e.g. Fairbanks, 1989; Hanebuth *et al.*, 2000; Yokoyama *et al.*, 2000). The disintegration of the great continental ice sheets of North America and Scandinavia, and also of the much smaller British-Irish ice sheet (BIIS), which together accounted for most of this increase in global sea level, resulted in significant changes in vertical crustal motion (Peltier, 1998). The mass of these continental ice sheets caused mantle deformation and crustal subsidence (glacioisostatic subsidence). Peripheral to these great ice sheets were areas of glacioisostatic uplift (creating a 'forebulge') resulting from the migration of mantle material from areas beneath the ice centres. The reduction in mass and eventual disappearance of these great ice sheets during the late glacial and early Holocene led to the glacioisostatic uplift of formerly glaciated land areas as the displaced mantle material returned. In contrast, glacioisostatic subsidence characterised areas peripheral to the former ice masses as mantle material migrated back towards areas formerly located beneath the ice-sheet centres (Clark *et al.*, 1978). Although the major changes in mass distribution occurred between *c.*21 and *c.*7 ka BP, the high viscosity of mantle material means that the solid Earth is yet to reach isostatic equilibrium and GIA processes continue.

GIA models have three inputs: an Earth model, an ice model and a model of ocean-water distribution associated with mass redistribution. The Earth model simulates the solid Earth response to loading and unloading. Several different Earth models exist incorporating a range of estimates for lithospheric thickness and the vertical profile of mantle viscosity (e.g. Lambeck, 1995; Peltier *et al.*, 2002; Peltier, 2004). The ice model simulates global changes in ice volume. Again, different global ice models exist (e.g. Fleming *et al.*, 1998; Bassett *et al.*, 2005) reflecting the uncertainty in the melting histories of different ice masses. Ice models are constrained by geological RSL records from far-field regions (e.g. Fairbanks, 1989; Hanebuth *et al.*, 2000). RSL records from far-field locations will most closely approximate ice-melt equivalent sea-level change because less isostatic adjustment is experienced in these areas. The change in sea level over the world's oceans arising from changes in global ice volume can be predicted by solving the sea-level equation (Farrell and Clark, 1976; Mitrovica and Milne, 2003), which is the third GIA model input. Importantly, the equation takes into account perturbations in the ocean surface resulting from changes in the Earth's gravity field (geopotential), as surface mass is redistributed and the Earth's rotational potential changes.

Because RSL is defined by the height of the ocean surface relative to the solid Earth (or ocean floor), GIA models can be used to predict changes in RSL. Any change in the height of the ocean surface and/or in the height of the solid earth's surface will result in a change in RSL. Therefore, Late Quaternary RSL records represent an integrated signal of ocean water volume, the redistribution of water within the ocean basins, and vertical movements of the continental and oceanic crust. The pattern of Late Quaternary RSL change differs significantly between regions that were once far removed from (far-field), peripheral to (intermediate-field), or located within (near-field) the former margins of the large Late Pleistocene ice sheets (Clark *et al.*, 1978). Glacioisostatic effects dominate the near-field RSL records, which show significant lowering throughout the Late Quaternary (e.g. Greenland; Long *et al.*, 2003). Far-field regions are dominated by hydroisostatic processes, but are also affected indirectly by glacioisostatic processes. Forebulge collapse in near/intermediate regions led to a process of 'ocean siphoning' (Mitrovica and Peltier, 1991), whereby Late Holocene RSL in far-field equatorial ocean basins fell as water migrated to fill the space created by the collapsing forebulge in ocean basins peripheral to the previously glaciated regions (Mitrovica and Milne, 2002).

RSL change can be estimated for a geographical location using the following model (Shennan *et al.*, 2006):

$$\Delta\xi_{rsl}\,(T,\varphi) = \Delta\xi_{eus}\,(T) + \Delta\xi_{iso}\,(T,\varphi) + \Delta\xi_{tect}\,(T,\varphi) + \Delta\xi_{local}\,(T,\varphi) \tag{1}$$

This equation demonstrates that a change in RSL ($\Delta\xi_{rsl}$) at a given geographical location (φ) and time (T) will result if there is a change in time-dependent ice-melt equivalent sea level ($\Delta\xi_{eus}\,(T)$), any isostatic movements ($\Delta\xi_{iso}\,(T,\varphi)$), any tectonic changes ($\Delta\xi_{tect}\,(T,\varphi)$) and any local changes ($\Delta\xi_{local}\,(T,\varphi)$), such as land subsidence due to sediment compaction. Using this model, RSL may be estimated if the input parameters of ice-melt equivalent sea level, isostatic, tectonic and local changes are known. This is not the case because, as discussed above, different Earth and ice models exist. Moreover, the local model input parameter can be particularly difficult to quantify (discussed below). The accuracy of the GIA model is highly dependent on the accuracy of the input parameters. The performance of GIA models is determined by how well they replicate reconstructed estimates of Late Quaternary RSLs based on geological data (often referred to as 'observed' RSL). A poor fit between observed and modelled RSL for a given location would reveal potential inaccuracies in the input parameters. Modifying model inputs to produce a better fit between model predictions and observations leads to a refinement of our knowledge of the input parameters (ice-sheet dimensions, meltwater flux, lithospheric thickness and mantle viscosity structure), and this is one of the fundamental strengths of modelling.

Peltier (1998) believes the British Isles to be the most exotic geographical region on Earth in terms of GIA and, consequently, of RSL change. This is attributed to the interaction of post-glacial ice-melt equivalent sea-level change and glacio-isostatic adjustment associated with peripheral forebulge collapse in the south and crustal uplift in the north resulting from the melting of the BIIS (Lambeck, 1991). For example, RSL records from Arisaig in north-west Scotland (Figure 9.1) are highly

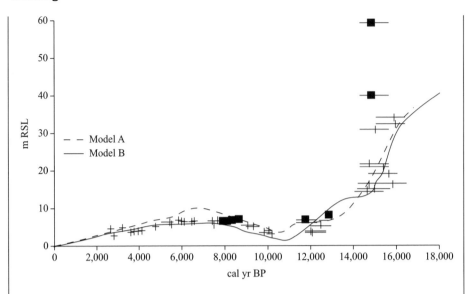

Figure 9.1 Comparison of model-predicted and observed RSL change at Arisaig, NW Scotland, from 16 ka BP to the present day. Reconstructed age and altitude uncertainty of each observed sea-level index point is represented by the horizontal and vertical error bar respectively. Relative sea level is located at an elevation below the limiting relative sea-level index points (represented by ■). Selection of different ice and earth model inputs results in different model predictions of RSL (Model A is based on Lambeck 1995, Model B is based on Peltier *et al.*, 2002). Reproduced with permission from Shennan, I., Hamilton, S., Hillier, C., Woodroffe, S. (2005) A 16,000-year record of near-field relative sea-level changes, northwest Scotland, United Kingdom. *Quaternary International* **133–134**, 95–106.

non-monotonic, showing a fall in RSL between *c.*16 ka BP and *c.*12 ka BP and after *c.*7.5 ka BP, associated with glacioisostatic uplift, interrupted by a period of RSL rise (*c.*10 ka BP–*c.*7.5 ka BP) as the rate of ice-melt equivalent sea-level rise outstripped the rate of glacioisostatic rebound (Shennan *et al.*, 2005). This is in contrast to the monotonic RSL records from southern England, which show continued submergence associated with ice-melt equivalent sea-level rise and isostatic subsidence related to the collapse of the Fennoscandian forebulge and the disintegration of the BIIS to the north (e.g. Gehrels *et al.*, 2011). The British Isles contains an extensive database of Late Quaternary RSL observations that can be used to constrain GIA models (Shennan *et al.*, 2006; Brooks *et al.*, 2008). These two factors (GIA spatial and temporal non-uniformity and an extensive database of empirical RSL estimates) have resulted in the British Isles being a fertile test bed for generations of GIA models (e.g. Lambeck, 1991; Peltier *et al.*, 2002; Shennan *et al.*, 2006).

Discrepancies between GIA model predictions and observed RSL data can highlight uncertainties in the accuracy of model inputs. For example, the British Isles GIA model of Lambeck (1993) appeared to underestimate the amount of rebound in northern Scotland (i.e. model predictions of RSL were much lower than that suggested by RSL observations in this area), perhaps resulting from the use of a local ice model that underestimated ice-sheet thickness over northern Scotland. In North Wales, different (post-2004) GIA model predictions of mid-Holocene RSL vary by up to 5 m (Roberts *et al.*, 2011), with some GIA models predicting a mid-Holocene highstand for which there is no evidence. For the most recent British Isles GIA model (Bradley *et al.*, 2011) an optimal fit between model estimates, Late Quaternary RSL observations and GPS measurements of present-day vertical land motion (Bradley *et al.*, 2009) is reached with a lithospheric thickness of 71 km, and upper and lower mantle viscosities of $4-6 \times 10^{20}$ Pa s and $\geq 3 \times 10^{22}$ Pa s, respectively. In this model, adoption of the global ice model of Bassett *et al.* (2005) produces a predicted mid-Holocene RSL highstand in North Wales. The Bassett *et al.* (2005) ice model is characterised by a rapid termination of melting (and hence of sea-level rise) during the mid-Holocene. However, the nature of mid- to late-Holocene ice-melt equivalent sea-level change is very uncertain because of the few RSL observations from far-field sites during this interval. By modifying the Holocene component of the ice model of Bassett *et al.* (2005), and assuming that ice-melt equivalent sea level continued to rise into the late Holocene, the predicted mid-Holocene highstand for North Wales is greatly reduced. A further refinement of the Bradley *et al.* (2011) British Isles GIA model has been attempted by using numerical ice flow models to constrain the size of the BIIS (Kuchar *et al.*, 2012). This is a departure from using a local ice model constrained by geomorphological evidence (Brooks *et al.*, 2008) and resulted in an increase in the estimated size (particularly in ice thickness) of the BIIS. An improved empirical data-model fit was achieved for some sites (e.g. Scottish highlands) but not others. This is consistent with the contention that trimlines, geomorphological features used in previous ice models to constrain ice-sheet thickness estimates, are indicators of minimum, rather than maximum, ice-surface height (Ballantyne, 2010).

Misfits between model estimates and observed estimates of RSL persist, which reveal that further refinement of Earth model and ice model inputs is necessary. An additional input model complexity, however, concerns the uncertainty often associated with local changes ($\Delta\xi_{local}$ (T,φ)). Local changes refer to any changes in the tidal regime ($\Delta\xi_{tide}$ (T,φ)) or in ground elevation resulting from the effects of sediment consolidation ($\Delta\xi_{sed}(T,\varphi)$). These particular model input parameters can be difficult to quantify. Tidal ranges were not constant through the Holocene (e.g. Gehrels *et al.*, 1995) but numerical model-based estimates of palaeotidal range require a detailed knowledge of changes in coastal geometry. This is only possible from comprehensive palaeogeographical reconstructions and is often beyond the scope of local sea-level reconstruction studies. Similarly, sediment compaction is an important, but poorly quantified, phenomenon. The position of a former RSL, known as a sea-level index point (SLI), can often be identified in the coastal deposits accumulating in tidal-dominated environments. SLIs from intercalated sediments are prone to sediment autocompaction, a process whereby the vertical thickness of unconsolidated sediments is reduced under self-weight. Autocompaction can lower SLIs from their original altitude of deposition and this can result in an overestimation of the rate and magnitude of RSL rise. The inevitable distortion of the original altitude of SLIs due to autocompaction is acknowledged but often not quantified in RSL studies, with age-altitude graphs of RSL portrayed that are usually uncorrected for autocompaction.

Only recently has the behaviour of lower intertidal sediments in response to compression been adequately investigated (Brain *et al.*, 2011). In some cases this problem can be overcome by using only 'basal' SLIs, where the sediment used to identify a former RSL is located directly above a substrate, such as glacial till, in which little or no sediment compression would have occurred (e.g. Long *et al.*, 2010). Nevertheless, an inability to accurately account for sediment compaction in model and observed RSL data sets complicates attempts to calibrate and refine GIA models using geological RSL data.

The precision of GIA models is increasing as a result of the iterative comparison with the growing database of empirical RSL estimates (which are themselves becoming more precise as the issues identified above are being addressed, e.g. Brain *et al.*, 2011). Therefore, so too is our knowledge about Late Pleistocene ice-sheet dynamics, global meltwater flux and Earth rheology. Beyond the application of GIA models in developing an understanding of Earth's response to Late Quaternary deglaciation, GIA models also have significant societal relevance. Coastal areas are amongst the most densely populated in the world; essential infrastructure, such as nuclear power stations, oil refineries and industrial facilities are concentrated in coastal locations, and an estimated \$1 trillion worth of assets is located <1 m above current sea level (Stern, 2007). The provision of maps detailing present-day spatial patterns of vertical land motion and sea-level change, generated using GIA models, is crucial for coastal managers and researchers involved in investigating potential hazards associated with future sea-level change (Shennan *et al.*, 2009).

Issues in modelling – bounding, parameters and uncertainty

The basic typology of models suggests that a number of key decisions need to be made by the researcher when deciding upon which model is most appropriate to represent reality. Building a model requires the same initial set of decisions as are required in building a system, as outlined in Chapter 7. The key initial decision is what to include and exclude from the model: where to draw the boundaries. The decision as to what to include in, or exclude from, a model defines the entities and relationships that will be modelled and explained. The definition of the environment of the model sets its context and what will not or cannot be modelled. Likewise, the inclusion of entities or variables within a model means that they need to be described and developed in a manner appropriate to the purposes of that model. Vegetation may be included in a fluvial model but it will be included in terms of the attributes thought to be important for the operation of the fluvial systems, not in the sense of a complete individual with all its attributes that make it a particular species and individual representative of that species. Similarly, people and their behaviour may be modelled in flood modelling but it is only their attributes as defined as important to the model by the modellers that will be considered. The behaviour of people as rounded individuals will not be considered as this will be viewed as irrelevant to the model. It is, however, these additional, irrelevant attributes that define and determine their behaviour outside of the context of the model. The existence of these unmodelled attributes may influence the model through their emotional attachment to their homes, their relatives and to the political pressure they exert to relieve flooding. Although this type of behaviour may not be modelled it can feedback into the reality the model is supposedly representing.

Within any model the entities defined need to be given values and the choice of these values can affect the operation of the model. It is important, therefore, to know and be able to justify the basis for these choices. The range of values that are acceptable for an entity in a model may reflect the experience of the modeller or constraints in the modelling process. Referring back to experience and pre-existing models to justify the use of values in current models inserts a circular set of arguments in the justification of modelling values. In the case of empirical-statistical models, for example, the parameters that define relationships are derived from pre-existing data sets as are the formal equations in the model (Demeritt and Wainwright, 2005).

Demeritt and Wainwright (2005) point out that a great amount of work in modelling is spent ensuring that parameters are defined and improved upon. Although the values of these parameters may be initially fixed by reference to the output of the original data set, over time with their application to other data sets the value of these parameters may alter or become optimised. The values of the parameters are 'fine-tuned' as more and more data sets are considered. Refinement of such parameters based on 'real' data is often referred to as 'physically based'. This process of fine-tuning can, however, limit the applicability of the model. The model may more precisely match the outputs of the empirical data set but these data are derived from a specific time and a specific place. This means that the fine-tuning and the parameter values it produces may be time and place specific. They may be excellent for 'explaining' variations in streamflow at that location for a particular time period but they are not as good at 'explaining' variations at other locations in other time periods. Although the variables and relationships might be the same or similar, the exact nature of these relationships might vary and so the parameters need to vary as well. Where the nature of the variable and relationships is different from place to place or over different time periods, then the parameters may not even be appropriate. In addition, the model is constructed and explained in terms of physical processes (usually). The model is then 'tweaked' until the parameters in it provide an output that matches the empirical data set. The parameters are assessed against a data set they have been tweaked to reflect. There is no way of knowing whether the physical model is correct and a valid explanation of the empirical data or whether the model is only correct because the parameters have been fine-tuned to produce an output that looks like the empirical data. The problem of circularity of argument is clear. The issue of parameterisation also raises the issue of the ability of models to explain in the face of the uniqueness of place. If a model is only calibrated for a specific location over a specific time period then is the same model transportable to another place and time?

Assessing the validity of models is a key concern for contemporary model builders in physical geography. Assuming that the model itself has no issues within its operation such as programming errors, the results produced reflect operation of reality according to the model. Matching the model outcomes to reality requires the modeller to express the outcomes in a form that is open to validation. Judging whether an outcome or outcomes match reality requires the modeller to define a set of criteria against which the outcome properties and the 'real' properties can be compared. The modeller will also need to decide a hierarchy of outcomes in terms of the importance of their match to reality, as some outcomes may match well, whilst others may have values beyond an acceptable level of difference. Selecting which outcomes are most significant in assessing a model may depend upon identifying the outcomes most directly related to the operation of processes within the model and so, by implication, those outcomes most closely associated with the causation the model is intended to simulate. A modeller may focus exclusively upon the properties produced within the model and so miss other, significant properties that are observable in reality. Likewise, the modeller will focus upon the observable properties that the model highlights as significant for the operation of the model.

This leads on to two other important issues. First, how close does the match have to be for the model to be validated? Second, even if a match can be identified how does the modeller know that the match to reality is for the reasons modelled? The matching of model outcomes to observations relies upon there being a clear correspondence or translation from model to some measurable property of reality. Even if a clear and justifiable translation exists, the issue of how 'close' the match in values needs to be before the model is validated needs to be clear. This decision is likely to be driven by the researcher(s) themselves, by the traditions and training in the subject as well as by the potential requirements of the models used.

Robustness in models is another issue that has generated a great deal of discussion (e.g. Levins, 1966; Orzack and Sober, 1993; Weisberg, 2004). Levins (1966) developed a simple definition of model robustness based on 'whether a result depends on the essentials of the model or on the details of the simplifying assumptions' (p. 20). Levins believed that by running a whole battery of models of a phenomenon with slightly different assumptions underlying each model it would be possible to assess the robustness of results. If the models produced similar results despite their differing assumptions then the results could be pronounced 'robust'. In this way Levins believed that what was trustworthy and valuable, as expressed by Wimsatt (1981), could be separated from what was unreliable and fleeting in a model. Robustness analysis is designed to separate the important from the illusory and accidental in models.

Although the general reception of the idea of assessing model robustness was favourable, some, such as Orzack and Sober (1993), viewed the practice as unscientific. They viewed robustness analysis as non-empirical confirmation of models and so a wholly inappropriate practice. They argue that the procedure proposed by Levins tested different models against each other. The models only vary in terms of their underlying assumptions and, possibly, their internal relationships. The empirical data that were used to guide model production, and against which a model is assessed, does not alter. The only thing being compared are the outputs from the slightly different models – there is no empirical basis for assessing which of the outputs is to be accepted, and so the whole procedure relies on researcher judgement rather than being grounded in the empirical data.

Weisberg (2004) thought that the issues surrounding model robustness, as discussed by Orzack and Sober (1993), were a result of a lack of clarity about the concept of 'robustness' as outlined by Levins. Weisberg views robustness as a four-step process. The first step involves the examination of a set of models to discover whether they all predict a similar result – the robust property that requires analysis. The second step involves an investigation of the structure of each model. Analysing how the model is put together and how this affects its operation can help to discover the cause or means by which the robust property is generated; it enables the researcher to identify the common structure of the models that results in the robust property. From these two initial steps it should be possible to state what the robust property of the set of models is and how the models produce this property and so how the two are linked. The third step is to translate the link into an empirical interpretation that then feeds into the fourth step, the testing of the link between common structures of the models and the robust property and the conditions under which this link could be severed. This last step identified the range of conditions under which the models will produce reliable results and those range of conditions for which they will not. In Weisberg terms, this last step identifies the '*ceteris paribus*' conditions for each set of models. In this way Weisberg was able to provide a clearer definition of what robustness meant and was able to ensure that robustness as a property was testable by reference to empirical data.

A key concern in modelling is the presence of uncertainty in environmental models. Uncertainty can take on a range of guises in modelling. Brown (2004) views uncertainty as being about the degree of belief or confidence a user has in the outcomes of a model. Uncertainties can arise at the initial stage of problem defining as well as ambiguities in the goals of the model itself. Such initial issues can then be compounded by uncertainty in structuring the model to match 'real' environmental relationships. Likewise, uncertainties in data generated by the model, and in the data derived as the basis for empirical testing, add a further layer of doubt to the model. Often the analysis of uncertainty is carried out as an 'end of pipe' procedure (Refsgaard *et al.*, 2007) much in the same manner as robustness, but Refsgaard *et al.* (2005) point out the importance of the analysis of uncertainty as being an ongoing process in model building. This means that the model builder should be aware of the sources and potential impacts of uncertainty from the point of defining the problem the model is built to analyse, through identifying the model objectives and in the construction of the model structure itself.

Brown (2004) suggests that there are three types of uncertainty (or certainty): certainty, bounded uncertainty and unbounded uncertainty. Bounded uncertainty refers to all possible outcomes being known and unbounded uncertainty refers to not all the outcomes being known. Within modelling it is only those outcomes associated with bounded uncertainty that can be included in the models. This is because probabilities need to be associated with each uncertainty so that each uncertainty needs to be known and related to the other known uncertainties to work out relative probabilities. Refsgaard *et al.* (2007) suggest that scenario analysis can be used when probabilities (rather than outcomes) are not known.

Walker *et al.* (2003) identifies five courses of uncertainty: context and framing, input uncertainty, model structure uncertainty, parameter uncertainty and model technical uncertainty. Context and framing refers to the bounding of the system that the model represents and can include the economic, environmental, social and technological conditions of the problem that has driven the production of the model. Input uncertainty refers to the uncertainties associated with the data used to drive the model. Model structure uncertainty refers to the uncertainty inherent in the conceptual construction of the model, its variables and relationships. Parameter uncertainty refers to, as mentioned above, the issues associated with parameter identification and quantification. Model technical uncertainty refers to the uncertainties associated with the technicalities of running the model on a computer including approximations required, space and time issues and software errors.

Refsgaard *et al.* (2007) suggest that uncertainty could be viewed as an uncertainty matrix. This provides a framework for defining and identifying different types of uncertainty associated with model building. The sources of uncertainty are context, inputs, model and model outputs. The five sources are roughly comparable to those identified by Walker *et al.* (2003), although it should be noted that the sources are not necessarily mutually exclusive and interactions between sources are likely to occur. These sources of error are manifested differently depending on the types of uncertainty and the nature of the uncertainty. Types of uncertainty are statistical, scenario, qualitative and recognised ignorance, whilst the natures of uncertainty are either epistemic or stochastic. Epistemic uncertainty arises due to imperfect knowledge about the nature of the reality and the system under study. Stochastic uncertainty (also referred to as ontological uncertainty by Refsgaard *et al.*, 2007) is the uncertainty due to the inherent variability of the reality under investigation.

Case Study

GLUE and modelling – analysing uncertainty

Beven and Freer (2001) outline the GLUE methodology in relation to modelling complex environmental systems. Beven (1993, 1996) is keen to avoid the conceptual issue of assuming that there is a single optimal model by highlighting the potential conceptual importance of equifinality. He defines this as: 'on grounds of physical theory there should be sufficient interactions among the components of a system that, unless the detailed characteristics of these components can be specified independently, many representations may be equally acceptable' (Beven and Freer 2001, pp. 11–12).

This definition accepts that uncertainty exists at all stages of the modelling process and that it is unreasonable to expect a single optimal model. Instead, there will be a set of model outcomes of the modelling process, all of which are potentially acceptable as 'valid' models with no single one being 'better' than the others. Equifinality provides the conceptual basis for generalised likelihood uncertainty estimation or GLUE (Beven and Binley, 1992). Within this assessment framework the researcher sets a range of parameter values based on their experience and previous knowledge. This set of values is used to generate independent random parameter sets for modelling. From these parameters a series of models are generated so producing the range of potential model outcomes for assessment. Assessment of each member of the series is based upon an analysis of the modelled residuals.

The process could be thought of as finding the real, highest peak on a landscape. The single outcome or highest peak represents 'reality'. It could also represent the single outcome that would be produced from a fully and effectively optimised model. This is the ideal that is the target of modelling although whether the two peaks coincide or not is another matter. GLUE produces the other, lower peaks in the landscape. These represent outcomes from the variations in parameterisation, each has a different location within the phase space and height representing how closely these outcomes match the idealised single optimal outcome. Sets of parameters are assessed rather than individual parameters so interactions between parameters are tested through this form of analysis as well.

The GLUE methodology was used to assess uncertainty in the modelled estimates of acid deposition across Wales (Page *et al.*, 2004). The Hull acid rainfall model (HARM), designed to model UK-wide estimates of acid deposition (HARM: Metcalfe *et al.*, 1995, 2001) was used to model acid deposition at 25 sites in Wales. The GLUE method was used to quantify the uncertainty associated with these modelled predictions. The HARM model has a set of parameters that require definition such as boundary layer depth, wind speed, dry deposition velocity, sulphur dioxide conversion rates and scavenging rates, as well as parameters associated with scaling of pollutant emissions. The key uncertainties in the modelling process seemed to be associated with emissions, particularly emissions inventories. Twenty-five emissions scenarios were selected to represent the emissions input parameter space. The outputs from these scenarios were compared with the 1995 acid deposition levels measured at 25 sites across Wales by the Welsh Acid Waters Survey (Stevens *et al.*, 1997), which was a smaller study than a similar 1984 survey of 44 sites (Donald and Stoner, 1989). These two surveys represented the best available data against which to assess the models. This highlights the issue that the assessment of uncertainty is only as good in quality as the data against which it can be assessed.

If all 25 sites were used in assessing the models, then all 100,000 simulations were rejected. Leaving aside four problematic sites meant that there were 2,101 parameter sets that modelled the behaviour of acid deposition across Wales. This means that these 2,101 parameter sets performed better than the rejection criteria set. If the 1984 survey data were used as the data to assess the models then a similar pattern of over and underestimation for acid deposition was found at common sites. This suggested that the model and the 2,101 parameter sets used produced a consistent interpretation of acid deposition across Wales.

Summary

A model is a simplification or an abstraction of reality. Modelling may employ a naïve pragmatic realism as its main conceptual approach as outlined by Beven. This approach places the model as an ever-improving representation of reality prevented from being a perfect reflection at the moment by the constraints of technology. Modelling based on 'reductionist' concepts insists that models can only be built from and understood in terms of the operation of 'fundamental principles'. 'Synthestists'-based modelling tries to model reality using entities, relationships and principles appropriate for that level of reality, recognising that reality is differentiated and stratified. This view of modelling is more aligned with the critical realist view of reality.

Different types of models can be found in use in physical geography ranging from analogue models through to deterministic models. Modelling requires consideration of key issues of bounding, variable and parameter identification, definition and quantification, place uniqueness, model validation, robustness and uncertainty. Assessing the level of uncertainty in modelling requires a clear understanding of the conceptual basis of modelling and the parameter-dependent nature of the outputs produced. In this respect, Beven's use of the concept of equifinality (through the GLUE methodology) to quantify the 'fit' of parameter sets to real data is an interesting approach.

Chapter 10

Physical geography and societies

Paradigms and social networks

The idea that scientific thought and its change is strongly influenced, if not determined, by the society within which it exists is something that Kuhn developed into the concept of paradigms. Although a paradigm is not a well-defined concept (there are at least 54 definitions according to Masterman, 1970), it has acted as a useful beacon for studies that emphasise the importance of social context for science and, probably more contentiously, for the idea that changes in scientific ideas are illogical and totally socially constructed. Even if you do not subscribe to the very strong social or even socially-determined view of change in scientific ideas, the development of science within a social vacuum cannot be sustained any longer. Science, and by implication physical geography, develops within a range of social networks. Even if social networks do not impact upon the logic of selecting theories or paradigms, they certainly contribute to what and how reality is studied.

Physical geographers are part of a society, in fact they are part of a number of inter-locking and interweaving societies. This detail is often lacking in a discussion of both paradigms and of the social construction of reality. The assumption of a single and easy to identify 'society' and, likewise, simple and singular 'social influences' is a fallacy that requires some correction. It may be more appropriate to think of any physical geographer (or indeed any scientist) in the context of their different social networks and their different sets of social relationships, which can influence how they perceive, study and, hence, understand reality. At the broadest scale, a physical geographer is a member of humanity. Humanity may impose upon them certain behavioural, emotional and even ethical constraints. This is rarely the level at which the term 'society' is aimed in studies of paradigms and social construction. A more common use of the term is in relation to membership of either Western or capitalist society. At this scale, the scientist is often viewed as some sort of automaton playing out (more often than not in a very negative and detrimental manner) a set of capitalistic and imperialist imperatives. This form of analysis has echoes of the simplistic analysis of imperial imperatives used in studying geography in the nineteenth century (e.g. Collier and Inkpen, 2002). Patterns of behaviour and study may be recognisable, but the subsuming of every physical geographer in the UK or USA as a direct and unthinking agent of capitalistic dogma, with every action and thought being explicable in such terms, is highly reductionist and overly simplistic. Even if actions and thoughts can be interpreted as being influenced by the individual being in a capitalist society, the other social networks of which the individual is a part, as well as the individual themselves, provide the framework within which such a broad social influence is mediated. It should not be expected, therefore, that classifying individuals as from the same society at this scale confers some sort of explanatory power onto that level. Any influence is mediated and negotiated though different social networks. As with the imperialist studies of geography, this can mean that different individuals can have very different ideas about what the imperatives of their societies are.

Moving down from the scale of capitalistic society, the problem of social influence upon scientific thought becomes more problematic and more interesting. An individual physical geographer is bound into social networks at different levels. The individual is a member of social networks defined by space, such as a nation or even a nation state, as well as more regional and local communities. At the smallest scale the individual has a tight and close social network of immediate work colleagues, friends and kin. Although often only the 'professional' relationships are highlighted in the analysis of social influences (e.g. Kuhn, 1962), other relationships may be equally as important. Any individual will be influenced by social networks other than professional ones, if only in terms of the amount of resources they devote to these other networks. An individual will also be located within seemingly aspatial social networks such as class and economic and international networks of fellow professionals. These are aspatial as they do not require space for their definition, although they do require a physical, and therefore spatial, manifestation for their operation. As the scale at which social networks are considered becomes increasingly refined, the detail required to understand the networks increases and the potential range of influences increases. The individual will respond and act within these different networks in different and not necessarily consistent ways. This makes the unravelling of a model of general social influences extremely difficult. The uniqueness of the individual and their interactions with their individually constructed social networks make such generalisations difficult and often irrelevant to understanding how ideas develop and change.

The above is not an argument for neglecting the social influences on the development of ideas in physical geography. Instead it is a plea for a more subtle and thoughtful approach to what the nature of this influence might be. The individual physical geographer should be placed in their complete and complex social networks rather than caricatured as unthinking and unfeeling passive agents of social trends. It is as responsive and emotive individuals that physical geographers interact within their different social networks, and indeed with different physical environments, and it is from this interaction that their beliefs and opinions concerning concepts in physical geography develop.

Despite the above comments, some general models of the influence of social networks on physical geographers and the development and selection of ideas have been developed. Stoddart (1981) provided a very general model of the hierarchical nature of the social networks within which physical geographers are embedded. The important aspect of the model is how the individuals themselves mediate the social influences. Not all social influences are the same however. Some social practices may not be viewed as social practices that have been constructed as they have become ingrained within their social networks as expected norms or rules of behaviour. This is equivalent to the 'disciplinary matrix' identified by Kuhn. It is the tangibles, such as textbooks and written documentation, and the intangibles, expected norms of behaviour, that make up the professional practice of 'doing' physical geography. The rules defined by such practices could refer to the type of questions to be asked, the methods used to answer particular questions, and the forms of analysis employed. These practices are not questioned and once established by continual practice and repeated publication, they become viewed as the way to undertake a particular type of study or the route to understanding a particular phenomenon.

Social construction and physical geography?

Recent discussions in geography have focused on the social construction of reality. Papers by Demeritt (1996, 2001) in relation to climatology have put forward the argument that the reality studied by physical geographers is socially constructed. Countering this view, Schneider (2001) has suggested that such a strong social constructive view is detrimental to the subject. Specifically, Deremitt highlights the relative nature of knowledge and denies the possibility of achieving an objective knowledge of the physical environment. Whilst social construction of reality is undeniable in the sense that physical geographers always work within social contexts, the strong 'sociology of science' view that particular brands of social construction demand, paints a relativistic picture of knowledge that most, if not all, physical geographers feel uncomfortable with.

Discussion of social constructionism should begin, as Hacking notes, with assessing why you need to assert something is socially constructed? Social construction in relation to social phenomenon usually develops an argument concerning the negative nature of the phenomenon under investigation. Central to the argument is the seemingly natural nature of the phenomenon despite its social construction. There is no discussion of the reality of the phenomenon; its negative nature is to be negated by its social construction and, by implication, its non-existence in an objective reality. Applying the same argument to physical geography implies that a physical and real basis to the entities and properties studied does not exist. As noted in Chapter 2, this view of reality is at odds with other philosophies such as critical rationalism, critical realism and pragmaticism, which all assume the existence of an independent reality. Although these philosophies do not assume that this reality can be known directly, they do assume that it exists and that physical geography is striving to explain it. Social constructivism, in its strongest sense, appears to deny this possibility or even encourage any attempt towards it.

It may be that the subject matter that is the focus of physical geography is different in nature from the subject matter of human geography. This is not to say that social constructivism in the 'weak' sense or limited sense identified above cannot be applied to the subject matter, but that the presence of an underlying reality that the subject matter refers to is not denied. As noted in Chapter 3, Hacking identified two types of kinds that can be the focus of study. Indifferent kinds are kinds or entities that do not react to the process of being classified. Interactive kinds are kinds or entities that do react and can respond to the process of being classified. Physical geography tends to deal with indifferent kinds or entities. They reflect the dialogue between the researcher and an independent reality. The indifferent kinds do not necessarily accurately mirror reality, but they do arise from a socially constructed, yet consistent, dialogue. It is this consistency and its agreed nature that provides the products of the dialogue with their objective power. Objective in this sense does not mean that the entities or kinds are 'real', only that they have been derived by a consistent process of dialogue and are treated as if real. The exact nature of the entity or kind may alter as the dialogue changes, but once defined the entities or kinds themselves influence the dialogue. In this sense it is pointless to begin to contrast 'real' as opposed to 'socially' constructed phenomenon. The dichotomy is a false one. In effect, phenomena are both real and social at the same time.

Ethics in physical geography – reflection required?

Ethical considerations in physical geography have not received much attention within the printed literature, although some debates have arisen at conferences. Scientific study in fields such as medicine and sociology has developed increasingly sophisticated means of assessing the ethical context of their subject matter. Given that humans are the focus of medical and sociological research this may seem an inappropriate place to look for guidelines about how to tackle ethical issues in physical geography. The context that these fields have established does, however, provide information on the type of questions that should be considered. Within physical geography there are specific areas that should be questioned. First, at an abstract level the acceptance of an external, independent reality imposes a series of moral obligations on the scientific researcher. If the researcher is committed to uncovering, as far as they can, the operation of this reality, then there is an implicit obligation to 'truth'. This is not as simple as it appears. Most physical geographers might assume that this means that the researchers should follow established procedures diligently and record (and revel in) success and failure equally. In other words, the researcher's work should reflect reality; the researcher holds a mirror up to reality and records the reflection without distortion. Hopefully, all of what has gone before illustrates the impossibility of such a naïve view of reality and its scientific study. Commitment to truth could be interpreted as ensuring that you derive data that reflects how reality is. Your view will be driven by theory and so truth and your theory becomes one and the same thing to you. Denying the theory is to deny the truth of how reality is. Reality may be interpreted to fit your theory. Ethically the researcher is committed to the truth, but other researchers may deny their version of the truth.

Commitment to the truth could be interpreted in another way. It could be viewed as commitment to use a specific approach to understanding reality and to abide by the outcomes of this approach. The centrality of falsification and testability in the philosophies outlined previously provides another means of reflecting reality. Researchers admit to the fallibility of their views of the world. Researchers submit their theories to rigorous testing in a dialogue with reality and with their co-researchers. Commitment to the truth then becomes commitment to this process of scientific scrutiny. It is through this method of rejection of theory that reality is to become knowable. Adherence to this process does, however, rely on an agreement between the researchers that the process of scrutiny is 'fair' and legitimate. Mechanisms for legitimising the process have evolved with the subject, from initial acceptance of the accounts of 'gentlemen' to the complicated peer-review systems for academic papers and research projects. The development of legitimate pathways does, however, also constrain research to that which is acceptable or capable of judgement by the formats available and specified by these pathways.

A further issue is that physical geography is embedded within social networks, as noted above. These networks include moral ones that are variable in space and time. These networks may dictate what constitutes appropriate sets of questions to ask, appropriate methods of scientific investigation, and appropriate forms of explanation. The early history of the study of the physical environment is littered with what seem now to be irrelevant moral constraints. Chorley *et al.* (1969) illustrate the need to retain the study of landscape development within a temporal framework that respected biblical events in the seventeenth and eighteenth centuries. The development of a catastrophic view of landscape development then produced a reaction of strict uniformitarianism within geological and evolutionary thinking. Within contemporary society, of which physical geographers are a part, there is increasing discussion of the moral rights of the environment and its components. Issues of preservation and conservation of the physical environment impinge directly on the subject matters studied by physical geographers. The indifferent entities and kinds that physical geographers study are part of the physical environment. Rivers, rocks and soils all contribute to the landscape that has become a focus for moral concerns both in its own right and as the basis for supporting ecosystems and particular rare species. Concern that these entities are preserved for future generations as they are now, or conserved as dynamic entities, is the basis for this moral concern. A whole literature has developed concerning moral and legal rights of future generations. Likewise, this concern has resulted in the development of sets of principles by which research activity is judged in terms of its impact upon the physical environment.

Principles such as minimum impact are used within the US National Park Service to guide the assessment of the environmental impact of scientific research. These principles define the types of practical steps researchers need to build into their experimental designs to be granted permission to work within national parks. Principles such as minimum impact, for example, ask the researcher to consider how they reach the site. Do they take a car? Moreover, what damage will they cause by their interference? And so on. The minimum tool principle likewise requires the researcher to justify data collection using the proposed method. Issues such as environmental damage using a particular instrument relative to other instruments are near the top of the agenda. Each principle and its practical implications are designed with resource conservation in mind. The principles begin to question the assumptions researchers make about the appropriate methods of study and bring to bear issues of general responsibility for actions that may have rarely entered into a researcher's consideration before.

A more general view could be taken of the ethical considerations within physical geography. The practice of physical geography involves a range of ethical choices, some of which are not viewed as choices at all, but as established working practices. In a similar vein there has recently been concern for the 'invisible' researchers within geography (Laoire and Shelton, 2003). The Research Assessment Exercise (now Research Excellence Framework) in the UK has thrown into sharp focus the role and lack of career structure for post-doctoral researchers and even Ph.D. students. The trends in the employment of this group could be viewed in the same light as the general trend towards casualisation of the labour force as a whole with all the associated problems. Academics, often the employers of these individuals, have been relatively slow to appreciate and develop strategies for career development. Often academics working with short-term contract researchers do not see it as part of their job to think about their employees. Often the employees are unaware of what their rights are. Likewise, comments are usually made about the increased flexibility of work practices for researchers by full-time academics. The dialogue between the groups, if indeed there are identifiable groupings, is one of an unequal power relationship. Oddly, academics have rarely researched this personal and close relationship between themselves and their researchers. Increasingly, in the UK at least, as research funds become concentrated in a few 'top' institutions, the development of academic careers will be controlled by fewer and fewer individuals. The ethical implications of this have rarely, if ever, been discussed.

The above discussions illustrate the important role that issues of responsibility play in developing an ethical code for research in physical geography. Although policy instruments such as minimum impact and minimum tool impose a framework of compliance upon researchers, they can also be viewed as practical and codified practices for environmental responsibility. The development of an ethical framework can be imposed from above as in this case, but it is more relevant to ask why the framework has to be imposed in the first place. Questions of moral uncertainty may be part of the answer, but it is likely that unaddressed issues of researcher responsibility are also part of the problem. Scientific research tends to focus on a particular problem. As the book has highlighted and researchers such as Medawar have commented, science is the art of the possible. Researchers break questions down into manageable chunks and answer what they believe they can, using the tools that they have learnt to use. Questioning of the questions being asked and the impact of their investigations upon the state of the physical environment for future generations is often not considered during the framing of their research. Particularly when the research is focused around the short-term research grant, with its associated promotional rewards, individual justification for an action may outweigh any potential ethical concerns if they are even articulated. Development of an ethical framework needs to be developed within the individual researcher. A researcher may claim they have an ethical commitment to discover the 'truth' of reality, but what do they do if the method they intend to employ causes irreversible damage to the environment they are studying? Which ethical concern has the greater moral imperative? Development of such an individual ethical framework will require a lot more work at the level of individuals and in the training that they receive, as well as being translated as a payback in career success. If the context of the individual researcher is not given as much weight as the ethical argument, then the success of any ethical framework will be vulnerable to the personal practicalities of the individual.

Although great consideration has been given to the ethics of research in the physical environment, the same consideration should be given to ethics in the process of research. Increasingly, research is a group or team activity rather than the solitary undertakings of a lone scholar. Similarly, with the advent of the Research Assessment Exercises (now Research Excellence Framework) in the UK, as well as the traditional peer-review system for publications, success of research activity is dependent upon a community. How individuals deal with other individuals in this research process is as much the sphere of ethical considerations as is the output from the process – the research itself. Once again, this sort of issue is something that academics have not had to deal with and often does not even register as part of their ethical framework. Within the UK for example, there has recently been a concern over the casualisation of researchers, almost the development of a research underclass. The status of these individuals in the research process is often left to the vagaries of individual research team leaders. The responsibility of employer to employee, and vice versa, are part of this relationship, but one which academics are rarely trained for or necessarily have an ethical framework in which to assess. It is telling that a number of universities have developed rules for inclusion and precedence of authorship on academic papers involving such teams. These raise issues of unequal power relationships within the research process. The inequality of these relationships requires recognition and consideration that has been lacking in the recent past.

Case Study

Fraud in physical geography?

Goodstein (2010) suggests that there are five reasons for fraud, the faking or fabrication of data, being committed by academic and non-economic geologists (geologists involved in economic activities are assumed to succumb to old-fashioned greed when they fake data such as salting mines). The five reasons are: career pressure, laziness, the ability and power trip associated with 'getting away with it', financial gain (just like economic geologists), and ideology. The classic case of fraud in the earth sciences could be that unmasked by Talent (1989) of VJ Gupta and his supposedly Himalayan fossils.

Below are two cases of potential fraud that illustrate that there is not necessarily a straightforward way to identify fraud, and then to raise the issue of fraud.

Sabbagh (1999) outlined the fraud committed by Harrison, a professor in botany, in the 1930s to 1950s. The alleged fraud was the 'finding' of plants in the Isle of Rum, which seemed to confirm Harrison's theory that parts of the Hebrides had remained ice-free during the Devensian glacial stage. Sabbagh's book takes up the story of John Raven who investigated the startling theory in 1948 by visiting the area the plants had supposedly been found in order to try to uncover similar and collaborative evidence. Raven failed to find such evidence and submitted a report on his findings to the university. Sabbagh's key argument is that this report was not acted upon mainly due to the cultural context of the 1950s and the perceived need to ensure that there was not a scientific scandal attached to the university or to the professor. Evidence was judged secondary to social and cultural imperatives.

A similar point is made by Hooper (2002) in relation to the work on industrial melanism by Kettlewell (1952, 1955, 1956 and 1973). Industrial melanism is the rapid rise in number of dark forms on moth species near to manufacturing centres after the industrial revolution. The melanism is hypothesised to be the result of an increase in atmospheric pollution resulting in evolutionary changes that produce colour changes in moths so that they are camouflaged against the new, darker surfaces. Light-coloured moths are preferentially eaten by birds leaving the darker-coloured moths to reproduce. Although the methods and interpretations of Kettlewell have been critically reviewed (Majerus, 1998), Hooper contends that Kettewell 'fudged' his data to obtain the results he needed to confirm his theory. Hooper's argument rests upon, first, Kettlewell's notebooks not being available for analysis, having been 'lost' or destroyed, as she notes Kettlewell regularly did with his correspondence and research material (a claim rejected by Rudge, 2005). Second, Hooper notes that Kettlewell reports several false starts to his research on industrial melanism (Kettlewell, 1955) that she interprets as evidence that Kettlewell 'tinkered' with his experiments and data until he got the 'right' outcome. Finally, Hooper notes discrepancies between the recapture rates Kettlewell reports in two papers (Kettlewell, 1955, 1956). She concludes that the dramatic rise in the middle of the experiment was a result of Kettlewell 'fudging' his data to please the researchers in the Oxford School of Ecological Genetics (an alternative hypothesis of a change in weather conditions was rejected).

Rudge (2005) provides an interesting analysis of Hooper's arguments. Hooper's initial discussion of Kettlewell is concerned with his character and his association with the Oxford School of Ecological Genetics (Turner, 1985). Rudge suggests the intention is to show that Kettlewell was of dubious character and so someone who could be imagined to indulge in scientific fraud. His conviction in the existence of industrial melanism may have resulted in some over-interpretation of results but this could be said of a number of scientists convinced of their theories. Science operates by others repeating and falsifying such work, however, so 'wrong' theories can be corrected. Hooper contends that Kettlewell had motive in carrying out fraud: advancement of his career and advancing his standing within the Oxford School of Ecological Genetics. Rudge notes that Kettlewell's work was outside of the key period of 'evolutionary synthesis' to which the Oxford School of Ecological Genetics made a major contribution. Overall, Rudge judges that Kettlewell seemed to operate as a 'normal' scientist in his social and cultural context and that judging him on the standards of a different time and context is inappropriate. In addition, using a model of science in which the 'truth' is established in a single, specific way that allows for no learning from field experience, that allows only a single interpretation of the results, and that misses the intersubjective nature of scientific debate, may not be the most appropriate way to assess scientists.

More significantly, Rudge draws attention to the model of science that Hooper seems to be using to judge Kettlewell's work. In a field-based, historical science like evolutionary biology, refining experimental fieldwork is not uncommon. In fact, the detail that Kettlewell provided about his 'false' starts implies that he was keen to document both unsuccessful and successful approaches. Likewise, in a field science there is no possibility of achieving findings in controlled conditions. Experimental results in a field science are always tentative, always limited in the space and time. Extending these results beyond the specifics of their context is always fraught with problems and limitations.

Physical and human geography – division or integration?

Recent discussions over the nature of geography often refer to the human/physical divide in the subject. Despite protests and pleading to the contrary, the divide is often seen as insurmountable, with some even claiming that geography has become effectively split into two specialisms (some even carry the subdivide amongst specialisms further). Massey (1999) tried to tease out the commonalities between the two parts of geography via a consideration of the role of entities. Raper and Livingstone (1995) and Raper (2000) confirm the potential for discussion as did the ensuing printed debate (Massey, 1999; Lane, 2001; Raper and Livingstone, 2001). Despite these overtures across the divide, there seems to be little real movement of positions so far.

This debate about the possibility of integrating physical and human geography is not a new one. A special issue of *Transactions of the Institute of British Geographers* in 1986 provided a series of short papers on different geographers' opinions on the likelihood of such common study. This is not the first time that such concerns and hopes have been expressed (e.g. in *Land and Man* (Sauer, 1956)). Several geographers have explicitly been concerned with developing concepts that would facilitate this integration, such as Carl Sauer and the concept of landscape. By the end of the twentieth century, the debate still continued without resolution.

Keeping the two sides apart are centrifugal forces that vary from place to place, but which collectively are likely to prevent the integration wished for by many. Academics, as noted above, are only human. Career paths and funding requirements mean that people will pursue courses of action profitable to them in their particular context. If funding councils do not recognise an integrative physical/human geography then why pursue it? There is no money in it, there are no rewards. This may sound cynical, but an academic life is short, the period available to make an impact even shorter, so why waste it in pursuit of an unattainable goal? There is also the feeling amongst key researchers on both sides that what they study is different from the other side. The subject matter of human and physical geography is seen as fundamentally different. Hacking's (1999) distinction between indifferent and interactive kinds may be taken by some to be a good description of the differences between physical and human geography respectively. This may be simplifying the matter, but it does have an appeal in explaining the continued divide. If the subject matter is different at a fundamental level then the techniques employed to study one kind cannot necessarily be easily transferred to the study of the other kind. This means a reductionist approach will not be of any use in a post-modern analysis of asylum seekers. Likewise, what could a postmodernist say of use to the study of gravel-bedded rivers? Acceptance of such a key divide will justify the lack of communication between the camps.

With such entrenched and well-funded camps is there any point in pursuing the issue of integrating physical and human geography? Although some may regard any integration as superficial and part of a propaganda drive to maintain a university subject that has run its course, there is belief amongst academics of the potential value of an integrative approach (e.g. Pitman, 2005). The methodologies of human geography can be applied to the study of the practice of physical geography. Collier and Inkpen

(2002, 2003) have, somewhat simplistically, begun to analyse how surveying at the Royal Geographical Society in the nineteenth century was influenced in its practice by the context of the period. Livingstone (2005) analyses Shaler's work in both eugenics and geology in the context of the development of Darwinian concepts at the time. Such historic studies can help us to see how seemingly objective facts and concepts are actually constructed by individuals in societies and how the nature of 'objectivity' is affected by these contexts.

Likely to be of more significance to current physical geography is the increasing recognition (stating the obvious maybe), that it is difficult to separate human and physical environments. The emergence of 'Earth System Science' recognises the equal importance of biophysical and social sciences in understanding the state and future of the Earth System. In the context of global warming, for example, we may understand the carbon cycle and be able to model the change in atmospheric thermal efficiency resulting from an increase in atmospheric CO_2. Nevertheless, the application of sophisticated climate models to predict future climate change relies on meaningful projections of greenhouse gas emissions, which can only be achieved by acquiring information on the potential changes in population, economic growth and development, and technological development. The interpenetration of the physical and human mean that it is difficult to justify that processes of environmental change are purely physical or that social structures rely solely upon human processes. Maintaining that one realm does not influence the other would be to view each as meshing together at increasingly smaller scales, but never actually overlapping or merging. Gould (2001) provides a similar image of the relationship between science and religion, labelling it as 'fractal' as the two spheres mixed, but never merged, irrespective of the scale of resolution they were observed at (Figure 10.1). In this case, the two never merged as each answered questions about a specific realm, science what reality was like, religion, how we ought to live. One could not answer questions about the other. Keeping human and physical geography separate implies the same key distinction. Interestingly, however, this separatist view would not actually deny the importance of either branch of geography; it would merely accept that human and physical questions were distinct and separate, but equally as valid to ask in the first place. This view would enable an uneasy truce to prevail in the integrate/split conflict.

An important observation that has rarely been considered is that there are distinct approaches within both the study of human and physical geography with differing methodologies and views of reality. Arguments for separation and even some for integration have usually focused upon the extreme members of both realms. The argument seems to run that if these end members cannot be integrated then none of physical and human geography can. It may be that these end members follow Gould's fractal description, but other parts of human and physical geography may be easier, or rather more appropriate, to integrate. The approaches and topics that ask the same questions about reality, or can inform the questions that each subdiscipline asks, may be more appropriate candidates for integration. This may be an area that, as environmental degradation continues, becomes a concern of necessity for geography as a subject for both its relevance and survival.

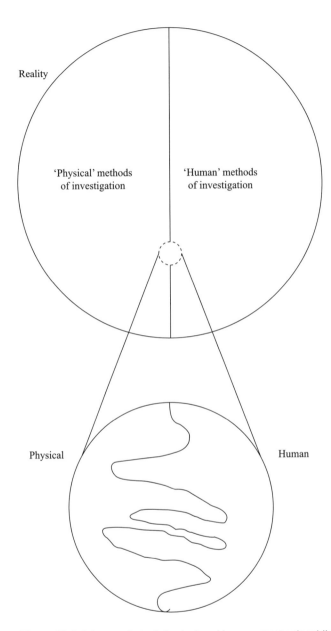

Figure 10.1 Interweaving of physical and human geography philosophies.

Participatory science and physical geography

An important development in the application of science recently has encouraged inter-disciplinary work (or multidisciplinary depending on your definitions), usually in teams, dragging human and physical geography together. These teams have often been drawn together around key projects or issues (cynically it might be argued for funding resources, less cynically it would be argued in order to provide a holistic approach to the problem). The advent of participatory science or democratic science, citizen science or community science (Lengwiler, 2008; Minkler, 2010; Harrison, 2011; Guehlstorf and Hallstrom, 2011; Jonsson *et al.*, 2011; Jacbosen *et al.*, 2012), or whatever term is coined, all point towards the same trend: the increasing involvement of non-expert knowledge in scientific analysis of the physical environment.

The involvement of the public and of 'lay' knowledge in science is not a new phenomenon. Blaikie's early work on soil erosion (1981, 1985) highlighted the important role that local or indigenous knowledge of soils made to the prevention of soil erosion. This was often contrasted with the imported 'expert' knowledge of soil erosion; this was heavily reliant on Anglo-American or European-based visions of poor practice, and expectations of agricultural land-use that were not necessarily in keeping with the 'novel' situations of subsistence-based and traditional agricultures. Within this context, however, the indigenous knowledge of soils is often seen as inferior to 'proper' scientific knowledge. Where indigenous knowledge or understanding of soil erosion is used, it has often been seen as a trivial addition to proper scientific understanding and it is only usable by that science if translated into a form that is coherent and consistent with the prevailing 'scientific' (usually Westernised view) of the 'real' problem. The dominance of the 'proper' scientific interpretation can be seen, for example, in the continuation of soil degradation policies based on its results by both India and China within the Himalayas. The continuation of the theory of Himalayan environmental degradation (THED; Blaikie and Muldavin, 2004), where the indigenous population and their practices are viewed as the problem, prevails despite contemporary research that contradicts its assumptions.

Participatory science is more than just playing 'lip-service' to being inclusive. Participatory science tends to refer to research done in collaboration by scientists and non-scientists, with no single group claiming 'knowledge' of the issue or problem under study. Knowledge evolves from the interaction and participation of the groups in the process of research. This means that although the scientists may have specific hypotheses they are pursing, the participation of non-scientists can radically alter the focus, scope and procedures of research. The engagement between the two (or more) groups is, therefore, both non-trivial and constructive. Communication, on both sides, is an essential aspect of such participatory approaches and devising a forum to enable the exchange of ideas, views and information is a key aspect of any project. Participation becomes a learning process for all involved and the outcomes from such projects cannot necessarily be judged by the scientific output alone.

Lane *et al.* (2010) highlight the work of Callon (1999) in thinking about the involvement of 'lay people' in science. Callon (1999) identifies three categories of involvement: the Public Education Model (PEM), the Public Debate Model (PDM) and the Co-Production of Knowledge Model (CKM). PEM identifies the case where scientific and lay knowledge are seen as distinct, different and, often, in opposition. Scientific knowledge is viewed as 'superior' and any interaction is firmly in the realms of the experts educating the lay. PDM shifts the basis of interaction between expert and lay. Scientific knowledge is still viewed as the force for generating knowledge about a problem, but the understanding gained is open to questioning or debate by those likely to be affected by its findings. Who is able to question and what can be questioned (e.g. the results of research, the process of research or even the assumptions of scientific research) are open to negotiation.

CKM is the ideal goal for participatory research. CKM does not accept that the knowledge is only generated by the expert or that this understanding is the only basis for debate. Knowledge is co-produced in a process that involves both expert and lay (however each category is internally differentiated), a process that is dynamic and, by implication, open to change on both sides. In this view of knowledge production, expertise is distributed across a whole range of people, both experts and lay. No single individual or approach can claim a privileged position or ownership of the knowledge co-produced; in fact, through the dynamic process a new and different type of knowledge about the problem is likely to result – one that could not have been predicted at the start of the process. Co-production could, for example, result in the reframing of the initial problem and so a re-analysis of the type of procedures needed to analyse the newly-defined issue. As noted by Lane *et al.* (2010) and Whatmore (2009), such a radical rethinking of issues can be unsettling for all parties as well as resulting in outcomes that are of uncertain value and unsettling to the status quo.

The example below of a participatory research project (based on analysing flood risk from Lane *et al.*, 2010) is a useful example of the approach researchers have taken in trying to develop a CKM. Such projects, however, still tend to be initiated by the experts. The funding for such projects is based on the perception of a problem by an organisation, government or a funding body, and then the funding of researchers to carry out the work, albeit in this case the research is to try to develop co-produced knowledge. The development of the Web and associated mapping and communication technologies has, however, given a glimpse of the possibility of 'bottom-up' or community-based, or even individually driven, participatory science. Community-based monitoring of environmental variables and issues has resulted in a number of community-run websites. Likewise, the monitoring of noise levels (e.g. Maisonneuve *et al.*, 2010), local pollution (e.g. Zwack *et al.*, 2011) and atmospheric pollution (e.g. Lahr and Kooistra, 2010) have all been collated, mapped and distributed by concerned communities. Such community-based projects provide an opportunity for collection of information at finer spatial and temporal resolutions than may be permissible in fund-restricted monitoring networks and projects. Similarly, a community can respond quickly to a specific event and provide 'on-the-spot' information on the development and impact of the event, often in terms not expected or conceptualised by scientists. It should be noted, however, that whenever discussing community-based projects, it is often a bit of a misnomer. Such projects tend to rely on a few determined, strong-willed individuals within communities, even though information collection may be distributed amongst a large number of individuals.

The scope for developing a 'grass-roots' community science is, however, likely to run up against a number of obstacles. Communities may be producing information but often concerns for an individual community form the basis for guiding the type of data collected. The issues that communities rally around and agree upon may reflect the prejudices and biases of that community and may slant the data collection and interpretation. Communities concerned about immigration might produce indices that reflect their beliefs in the problem or interpret relatively neutral information through the lens of their intolerance. Given the key role of 'leaders' in community-based projects, the potential for such rallying points should not be underestimated. Such community-derived information can reinforce, rather than aid, social justice. Communities may produce information on environmental issues, but what is the reliability of that information and, more importantly, how much belief do the different agencies involved in policymaking place on that information? Is the involvement of scientists still required within such projects in order to provide a veneer of respectability to the information and to elevate the material collected to the status of 'data'? Politicians and developers, for example, could quite rightly argue that any data produced by a community is biased in that it focuses on problems relevant to them at their local scale, and the community has much to gain if the information backs up their interests (although likewise the politicians and developers have a vested interest as well). Scientists have been viewed (when useful to whichever party is claiming 'objectivity' in their analysis) as impartial arbiters and creators of unbiased knowledge. Inclusion of scientists within a community-based project can enhance the belief in the information collected.

Conceptually, this issue of belief or confidence in the information collected could be viewed as a cone of certainty (or uncertainty depending on which end you start from). Take an example such as atmospheric pollution. A scientific network of monitoring sites could be seen as providing the basis of certainty (the broadest, lowest section of the cone). Community-based monitoring provides another level of information, usually further up the cone. The information from these surveys may be at a spatially higher resolution but it may also be intermittent, monitored by old equipment or by less precise methods. How should belief or confidence in this information be assessed? If the patterns or trends identified in the community-based information match those identified by the scientific data then it is suggestive that the two sets of information are triangulating onto a 'real' phenomenon. In this case, there may be increased belief or confidence in the community-based information and in the finer spatial resolution of the phenomenon it is implying. Similarly, scientists can also ensure that the information collected by a community is calibrated and open to rigorous sampling methods. Such a view, however, reinstates the privileged nature of 'scientific' and 'expert' knowledge, but this may be a necessity for securing funds for such projects, or for persuading policymakers of the benefits and accuracy of community-based information. There is an alternative view of the cone. Placing the scientific research at the base of the cone implies that this is the 'gold' standard for comparison – very much a scientist's view of belief. What happens if the community view is placed at the base of the cone and then information collected by other means is compared relative to it? Questions concerning the basis for this view are brought into sharper focus when considering the emotive concerns of the community and their detailed and spatially intimate local knowledge. The scientific layer now has to be viewed in terms of its connection to the community-derived information. How can the scientifically derived patterns be justified relative to the community-derived information? Is there a common language that can be negotiated or are the two communities talking across each other? Merging the two levels into a new, negotiated level of common agreement is akin to the co-production of knowledge outlined by Callon (1999) and Lane *et al.* (2010).

Summary

Physical geographers work within different societies at different scales. All are members of different networks and may behave in different, even contradictory, manners depending on the network being considered. Individuals should not be seen, however, as automatons responding blindly or consistently to their context. Context is mediated by individuals in a complicated and often unique manner. Community-based and community-driven research has increasingly been viewed as a way to develop socially meaningful research, but the integration, or rather negotiation, of the different knowledges involved is a difficult process involving loss of perceived power for all parties involved. Recently, ethical concerns have become an important consideration in physical geography, whether forced upon researchers through instruments such as minimum impact, or from within by moral considerations. There is little for physical geographers to follow in terms of a coherent and explicit professional code of conduct. Ethical obligations to the truth, to the physical environment being studied, to other researchers and to the individual, are all aspects of the research process that are not usually considered in detail.

References

Allen, T.F.H. and Starr, T.B. (1982) *Hierarchy: A perspective for ecological complexity.* University of Chicago Press, Chicago.

Alley, R.B. and Clark, P.U. (1999) The deglaciation of the Northern Hemiphere: A global perspective. *Annual Review of Earth and Planetary Sciences*, **27**, 149–182.

Alvarez, L.W., Alvarez, W., Asaro, F. and Michel, H.V. (1980) Extraterrestrial cause for the Cretaceous-Tertiary extinction. *Science*, **208**, 1095–1108.

Alvarez, W., Claeys, P. and Kieffer, S.W. (1995) Emplacement of Cretaceous-Tertiary boundary shocked quartz from Chicxulub crater. *Science*, **269**, 930–935.

Amspoker, M.C. and McIntire, D.C. (1978) Distribution of intertidal diatoms associated with sediments in Yaquina Estuary, Oregon. *Journal of Phycology*, **14**, 387–395.

An, Z., Porter, S.C., Zhou, W., Lu, Y., Donahue, D.J., Head, M.J., Wu, X., Ren, J. and Zheng, H. (1993) Episode of strengthened summer monsoon climate of Younger Dryas age on the loess plateau of Central China. *Quaternary Research*, **39**, 45–54.

Anderson, D.G., Goodyear, A.C., Kennett, J., and West, A. (2011) Multiple lines of evidence for possible Human population decline/settlement reorganization during the early Younger Dryas. *Quaternary International*, **242**, 570–583.

Andersson, L. (1996) An ontological dilemma: Epistemology and methodology of historical biogeography. *Journal of Biogeography*, **23**, 269–277.

Apol, M.E.F., Etienne, R.S. and Olff, H. (2008) Revisiting the evolutionary origin of allometric metabolic scaling in biology. *Functional Ecology*, **22**, 1070–1080.

Atkinson, T.C., Briffa, K.R. and Coope, G.R. (1987) Seasonal temperatures in Britain during the past 22,000 years, reconstructed using beetle remains. *Nature*, **325**, 587–592.

Bagnold, R.A. (1941). *The physics of blown sand and desert dune*s. Methuen & Co., London.

Baker, V.R. (1996a) The pragmatic roots of American Quaternary geology and geomorphology. *Geomorphology*, **16**, 197–215.

Baker, V.R. (1996b) Hypotheses and geomorphological reasoning. In B.L. Rhoads and C.E. Thorn (eds), *The scientific nature of geomorphology*. Wiley, Chichester, 57–85.

Baker, V. R. (1999) Geosemiosis. *Bulletin of the Geological Society of America*, **111**, 633–645.

Baker, V.R. and Twidale, C.R. (1991) The reenchantment of geomorphology. *Geomorphology*, **4**, 73–100.

Ballantyne, C.K. (2010) Extent and deglacial chronology of the last British–Irish Ice sheet: Implications of exposure dating using cosmogenic isotopes. *Journal of Quaternary Science*, **25**, 515–534.

Bassett, S.E., Milne, G.A., Mitrovica, J.X. and Clark, P.U. (2005) Ice sheet and solid earth influences on far-field sea-level histories. *Science*, **309**, 925–928.

Basu, P.K. (2003) Theory-ladenness of evidence: A case study from history of chemistry. *Studies in History and Philosophy of Science*, **34**, 351–368.

Battarbee, R.W., Flower, R.J., Stevenson, J. and Rippey, B. (1985) Lake acidification in Galloway: A palaeological test of competing hypotheses. *Nature*, **314**, 350–352.

Beatty, J. (1987) On behalf of the semantic view. *Biology and Philosophy*, **2**, 17–23.

Beninca, E., Huisman, J., Heerkloss, R., Jöhnk, K.D., Branco, P., Van Nes, E.H., Scheffer, M. and Ellner, S.P. (2008) Chaos in a long-term experiment with a plankton community. *Nature*, **451**, 822–825.

Benn, D.I. (1997) Glacier fluctuations in western Scotland. *Quaternary International*, **38–39**, 137–147.

Bennion, H. and Battarbee, R. (2007) The European Union Water Framework Directive: Opportunities for palaeolimnology. *Journal of Paleolimnology*, **38**, 285–295.

Bennion, H., Fluin, J. and Simpson, G.L. (2004) Assessing eutrophication and reference conditions for Scottish freshwater lochs using subfossil diatoms. *Journal of Applied Ecology*, **41**, 124–138.

Beven, K.J. (1993) Prophecy, reality and uncertainty in distributed hydrological modeling. *Advanced Water Resources*, **16**, 41–51.

Beven, K.J. (1996) Equifinality and uncertainty in geomorphological modeling. In B.L. Rhoads and C.E. Thorn (eds), *The scientific nature of geomorphology*. Wiley, Chichester, 289–313.

Beven, K. (2002) Towards a coherent philosophy for modelling the environment. *Proceedings of the Royal Society A: Mathematical, Physical and Engineering Sciences*, **458**, 2465–2484.

Beven, K.J. and Binley, A.M. (1992) The future of distributed models: Model calibration and uncertainty prediction. *Hydrological Processes*, **6**, 279–298.

Beven, K. and Freer, J. (2001) Equifinality, data assimilation, and uncertainty estimation in mechanistic modeling of complex environmental systems using the GLUE methodology. *Journal of Hydrology*, **249**, 11–29.

Bhaskar, R. (1978) *A realist theory of science*. Harvester Press, Brighton, Sussex.

Bhaskar, R. (1989) *Reclaiming reality: A critical introduction to contemporary philosophy*. Verso, London.

Bice, D.M., Newton, C.R., McCauley, S., Reiners, P.W. and McRoberts, C.A. (1992) Shocked quartz at the Triassic-Jurassic boundary in Italy. *Science*, **255**, 443–446.

Biron, P., Roy, A.G. and Best, J.L. (1996) Turbulent flow structure at concordant and discordant open-channel confluences. *Experiments in Fluids*, **21**, 437–446.

Bishop, P. (1980) Popper's principle of falsifiability and the irrefutability of the Davisian cycle. *Professional Geographer*, **32**, 310–315.

Björck, S., Walker, M.J.C., Cwynar, L.C., Johnsen, S., Knudsen, K-L., Lowe, J.J., Wohlfarth, B. and INTIMATE Members (1998) An event stratigraphy for the Last Termination in the North Atlantic region based on the Greenland ice-core record: A proposal by the INTIMATE group. *Journal of Quaternary Science*, **13**, 283–292.

Blaauw, M., Holliday, V.T., Gill, J.L. and Nicoll, K. (2012) Age models and the Younger Dryas impact hypothesis. *Proceedings of the National Academy of Sciences*, **109**, E2240.

Blaikie, P. (1981) Class, land use and soil erosion. *Development Policy Review*, **A14**, 55–77.

Blaikie, P. (1985) *The political economy of soil erosion*. Longman, London.

Blaikie, P.M. and Muldavin, J.S.S. (2004) Upstream, downstream, China, India: The politics of environment in the Himalayan region. *Annals of the Association of American Geographers*, **94**, 520–548.

Bogard, D.G. and Tiederman, W.G. (1986) Burst detection with single-point velocity measurements. *Journal of Fluid Mechanics*, **162**, 389–413.

Bogen, J. and Woodward, J. (1988) Saving the phenomena. *The Philosophical Review*, **97**, 303–356.

Bohor, B., Modreski, P.J. and Foord, E.E. (1987) Shocked quartz in the cretaceous-tertiary boundary clays: Evidence for a global distribution. *Science*, **236**, 705–709.

Bowler, P.J. (1983a) *The eclipse of Darwinism. Anti-Darwinian evolution theories in the decades around 1900*. Johns Hopkins University Press, Baltimore.

Bowler, P.J. (1983b) *Evolution: The history of an idea*. University of California Press, Berkeley.

Bradbrook, K.F., Biron, P., Lane, S.N., Richards, K.S. and Roy, A.G. (1998) Investigation of controls on secondary circulation and mixing processes in a simple confluence geometry using a three-dimensional numerical model. *Hydrological Processes*, **12**, 1371–1396.

Bradley, R.S. and England, J.H. (2008) The Younger Dryas and the sea of ancient ice. *Quaternary Research*, **70**, 1–10.

Bradley, S.L., Milne, G.A., Teferle, F.N., Bingley, R.M. and Orliac, E.J. (2009) Glacial isostatic adjustment of the British Isles: New constraints from GPS measurements of crustal motion. *Geophysical Journal International*, **178**, 14–22.

Bradley, S.L., Milne, G.A., Shennan, I. and Edwards, R. (2011) An improved Glacial Isostatic Adjustment model for the British Isles. *Journal of Quaternary Science*, **26**, 541–552.

Bradshaw, E.G., Nielsen, A.B. and Anderson, N.J. (2006) Using diatoms to assess the impacts of prehistoric, pre-industrial and modern land-use on Danish lakes. *Regional Environmental Change*, **6**, 17–24.

Brain, M.J., Long, A.J., Petley, D.N., Horton, B.P. and Allison, R.J. (2011) Compression behaviour of minerogenic low energy intertidal sediments. *Sedimentary Geology*, **233**, 28–41.

Broecker, W.S. (1992) The great ocean conveyor. In B.G. Levi, D. Hafemeister and R. Scribner (eds), *Global warming: Physics and facts*. AIP Conference Proceedings **247**. American Institute of Physics, New York, 129–161.

Broecker, W.S. (2006) Was the Younger Dryas triggered by a flood? *Science*, **312**, 1146–1148.

Broecker, W.S. and Denton, G.H. (1990) The role of ocean-atmosphere reorganizations in glacial cycles. *Quaternary Science Reviews*, **9**, 305–341.

Broecker, W.S., Peteet, D.M. and Rind, D. (1985) Does the ocean-atmosphere system have more than one stable mode of operation? *Nature*, **315**, 21–26.

Broecker, W.S., Kennett, J.P., Flower, B.P., Teller, J.T., Trumbore, S., Bonani, G. and Wolfi, W. (1989) Routing of meltwater from the Laurentide Ice Sheet during the Younger Dryas cold episode. *Nature*, **341**, 318–321.

Broecker, W.S., Denton, G.H., Edwards, R.L., Cheng, H., Alley, R.B. and Putnam, A.E. (2010) Putting the Younger Dryas cold event into context. *Quaternary Science Reviews*, **29**, 1078–1081.

Brooks, A.J., Bradley, S.L., Edwards, R.J., Milne, G.A., Horton, B. and Shennan, I. (2008) Postglacial relative sea-level observations from Ireland and their role in glacial rebound modelling. *Journal of Quaternary Science*, **23**, 175–192.

Brown, J.D. (2004) Knowledge, uncertainty and physical geography: Towards the development of methodologies for questioning belief. *Transactions of the Institute of British Geographers*, **29**, 367–381.

Brown, G.L. and Thomas, S.W. (1977) Large structure in a turbulent boundary layer. *Physics of Fluids*, **20**, s243–s252.

Brunsden, D. (1990) Tablets of stone: Toward the ten commandments of geomorphology. *Zeitschrift fur Geomorphologie*, **Suppl. Bd. 79**, 1–37.

Brunsden, D. (1993) The persistence of landforms. *Zeitschrift fur Geomorphologie*, **Suppl. Bd. 93**, 13–28.

Brunsden, D. (2001) A critical assessment of the sensitivity concept in geomorphology. *Catena*, **42**, 99–123.

Brunsden, D. and Thornes, J.B. (1979) Landscape sensitivity and change. *Transactions of the Institute of British Geographers*, **4**, 463–484.

Buffin-Belanger, T., Roy, A. and Kirkbride, A. (2000) On large-scale flow structures in a gravel-bed river. *Geomorphology*, **32**, 417–435.

Bunge, W. (1962). Theoretical geography. *Lund Studies in geography*. CWK Gleerup, Lund.

Callon, M. (1999) Actor-network theory – the market test. Blackwell, Oxford.

Carson, M.A. and Kirkby, M.J. (1972). *Hillslope form and process*. Cambridge University Press, Cambridge.

Cartwright, N. (1979) Causal laws and effective strategies. *Nous*, **13**, 419–437.

Cartwright, N. (1983) *How the laws of physics lie*. Oxford University Press, Oxford.

Cartwright, N. (1989) *Nature's capacities and their measurement*. Oxford University Press, Oxford.

Cartwright, N. (1999) *The dappled world*. Cambridge University Press, Cambridge.

Cartwright, N. (2003) Two theorems on invariance and causality. *Philosophy of Science*, **70**, 203–224.

Cartwright, N. (2007) *Hunting causes and using them*. Cambridge University Press, Cambridge.

Cartwright, N. and Efstathiou, S. (2011) Hunting causes and using them: Is there no bridge from here to there? *International Studies in the Philosophy of Science*, **25**, 223–241.

Chamberlin, T.C. (1890) The method of multiple working hypotheses. *Science* **XV**, 92–96.

Cheng, H., Edwards, R.L., Wang, Y., Kong, X., Ming, Y., Kelly, M.J., Wang, X., Gallup, C.D. and Liu, W. (2006) A penultimate glacial monsoon record from Hulu Cave and two-phase glacial terminations. *Geology*, **34**, 217–220.

Chorley, R.J. (1964) Geography and analogue models. *Annals of the Association of American Geographers*, **54**, 127–147.

Chorley, R.J. (1978) Bases for theory in geomorphology. In C. Embleton, D. Brunsden and D.K.C. Jones (eds), *Geomorphology: Present problems and future prospects*. Oxford University Press, Oxford, 1–13.

Chorley, R.J. and Haggett, P. (eds.) (1967) *Models in geography*. Methuen, London.

Chorley, R.J. and Kennedy, B.A. (1971) *Physical geography: A systems approach*. Prentice-Hall International, London.

Chorley, R.J., Beckinsale, R.P. and Dunn, A.J. (1969) *The history of the study of landforms or the development of geomorphology. Volume one*, Methuen, London.

Church, M. (2010) The trajectory of geomorphology. *Progress in Physical Geography*, **34**, 265–286.

Church, M., Hassan, M.A. and Wolcott, J.F. (1988) Stabilizing self-organized structures in gravel-bed stream channels: Field and experimental observations. *Water Resources Research*, **34**, 3169–3179.

Clark, W.C. and Dickson, N.M. (2003) Sustainability science: The emerging research programme. *Proceedings of the National Association of Science*, **100**, 8059–8061.

Clark, J.A., Farrell, W.E. and Peltier, R.W. (1978) Global changes in postglacial sea level: A numerical calculation. *Quaternary Research*, **9**, 265–287.

Clark, P.U., Licciardi, J.M., MacAyeal, D.R. and Jenson, J.W. (1996) Numerical reconstruction of a soft-bedded Laurentide Ice Sheet during the last glacial maximum. *Geology*, **24**, 679–682.

Clayden, B. (1982) Soil classification. In E.M. Bridges and D.A. Davidson (eds), *Principles and applications of soil geography*. Longman, London, 58–96.

Cleland, C.E. (2001) Historical science, experimental science and the scientific method. *Geology*, **29**, 987–990.

Cleland, C. (2002) Methodological and epistemic difference between historical science and experimental science. *Philosophy of Science*, **69**, 474–496.

Clifford, N. and Richards, K. (2005) Earth Science System: An oxymoron? *Earth Surface Processes and Landforms*, **30**, 378–383.

Cloke, P., Philo, C. and Sadler, D. (1991). *Approaching human geography: an introduction to contemporary theoretical debates*. Paul Chapman, London.

Cloos, H. (1949) *Gesprach mit der Erde*. Pijper and Co., Munich.

Collier, A. (1994) *Critical realism: An introduction to Roy Bhaskar's philosophy*. Verso, London.

Collier, P. and Inkpen, R. (2002) The RGS, exploration and the Empire and the contested nature of surveying, 1870–1914. *Area*, **34**, 273–283.

Collier, P. and Inkpen, R. (2003) The Royal Geographical Society and the development of surveying, 1870–1914. *Journal of Historical Geography*, **29**, 93–108.

Collier, P., Dewdeny, C., Inkpen, R., Mason, H. and Petley, D. (1999) On the non-necessity of quantum mechanics in geomorphology. *Transactions of the Institute of British Geographers*, **24**, 227–230.

Cooke, T.D. and Campbell, D.T. (1979) *Quasi-experimentation: Design and analysis issues for field settings*. Houghton Mifflin, Boston.

Cooke, R.U. (1984). Geomorphological hazards in Los Angeles; a study of slope and sediment problems in a metropolitan county. *The London Research Series in Geography 7*, Allen and Unwin, London.

Cooke, R.U. and Doornkamp, J. C. (1974). *Geomorphology in environmental management*. Oxford University Press, Oxford.

Cooke, R.U. and Warren, A. (1973). *Geomorphology in deserts*. BT Batsford Ltd., London.

Coones, P. (1987) *Mackinder's 'Scope and methods of geography' after a hundred years*. School of Geography, University of Oxford.

Cope, E.D. (1887) *The origin of the fittest*. Macmillian, New York.

Costantino, R.F., Desharnais, R.A., Cushing, J.M. and Dennis, B. (1997) Chaotic dynamics in an insect population. *Science*, **275**, 389–391.

Crozier, M.J., Vaughan, E.E. and Tippett, J.M. (1990) Relative instability of colluvium-filled bedrock depressions. *Earth Surface Processes and Landforms*, **15**, 326–339.

Dansgaard, W., White, J.W.C. and Johnsen, S.J. (1989) The abrupt termination of the Younger Dryas climate event. *Nature*, **339**, 532–533.

Davis, W.M. (1905) The geographical cycle in an arid climate. *Journal of Geology*, **13**, 381–407.

Dawkins, R. (1978) *The selfish gene*. Granada Publishing Limited, London.

De Boer, D.H. (1992) Hierarchies and spatial scale in process geomorphology: A review. *Geomorphology*, **4**, 303–318.

De Groot, R., Fisher, B., Christie, M., Aronson, J., Braat, L. and Gowdy, J. (2010) Integrating the ecological and economic dimensions in biodiversity and ecosystem service valuation. In P. Kumar (ed.), *The economics of ecosystems and biodiversity: Ecological and economic foundations*. Earthscan, London, 9–40.

Deely, J. (1990) *Basic of semiotics*. Indiana University Press, Bloomington.

Deevey, E.S. (1969) Coaxing history to conduct experiments. *BioScience*, **19**, 40–43.

DeLanda, M. (2006) *A new philosophy of society: Assemblage theory and social complexity*. Continuum, London.

Demeritt, D. (1996) Social theory and the reconstruction of science and geography. *Transactions of the Institute of British Geographers*, **21**, 484–503.

Demeritt, D. (2001) The construction of global warming and the politics of science. *Annals of the Association of American Geographers*, **91**, 307–337.

Demeritt, D. and Dyer, S. (2002) Dialogue, metaphors of dialogue and understandings of geography. *Area*, **34**, 229–241.

Demeritt, D. and Wainwright, J. (2005) Models, modelling, and geography. In N. Castree, A. Roger and D. Sherman (eds), *Questioning geography*. Blackwell, Oxford, 206–225.

Denton, G.H., Alley, R.B., Cormer, G.C. and Broecker, W.S. (2005) The role of seasonality in abrupt climate change. *Quaternary Science Reviews*, **24**, 1159–1182.

Derbyshire, E.D. (1973). *Climatic Geomorphology*. Longmans, London.

Dodds, P.S., Rothman, D.H. and Weitz, J.S. (2001) Re-examination of the '3/4-law' of metabolism. *Journal of Theoretical Biology*, **209**, 9–27.

Donald, A.P. and Stoner, J.H. (1989) The quality of atmospheric deposition in Wales. *Archives of Environmental Contamination and Toxicology*, **18**, 109–119.

Douglas, I. (1980); Climatic geomorphology. Present-day processes and landform evolution. Problems of interpretation. *Zeitschrift. fur Geomorphologie*, **36**, 27–47.

Dragovich, D. and Egan, M. (2011) Salt weathering and experimental desalination treatment of building sandstone, Sydney (Australia). *Environmental Earth Sciences*, **62**, 277–288.

Driver, F. (1992) Geography's empire: Histories of geographical knowledge. *Environment and Planning D: Society and Space*, **10**, 23–40.

Driver, F. (2000) Field-work in geography. *Transactions of the Institute of British Geographers*, **25**, 267–268.

Driver, F. (2001) Geography militant: Cultures of exploration and empire. Wiley-Blackwell, Oxford.

Dunne, J.A., Williams, R.J. and Martinez, N.D. (2002) Network structure and biodiversity loss in food webs: Robustness increases with connectance. *Ecology Letters*, **5**, 558–567.

Dupre, J. (1993) *The disorder of things: Metaphysical foundations of the disunity of science.* Harvard University Press, Cambridge, Mass.

Dupre, J. (2001) In defence of classification. *Studies in the history and philosophy of biology and biomedical science*, **32**, 203–219.

Elner, R.W. and Vadas, Sr., R.L. (1990) Inference in ecology: The sea urchin phenomenon in the northwestern Atlantic. *The American Naturalist*, **136**, 108–125.

Elsig, J., Schmitt, J., Leuenberger, D., Schneider, R., Eyer, M., Leuenberger, M., Joos, F., Fischer, H. and Stocker, T.F. (2009) Stable isotope constraints on Holocene carbon cycles changes from an Antarctic ice core. *Nature*, **461**, 507–514.

Embleton, C. and King, C.A.M. (1975a). *Periglacial geomorphology*. Second edition. John Wiley and Sons, New York.

Embleton, C. and King, C.A.M. (1975b). *Glacial geomorphology*. Second edition. John Wiley and Sons, New York.

Engstrom, D.R., Fritz, S.C., Almendinger, J.E. and Juggins, S. (2000) Chemical and biological trends during lake evolution in recently deglaciated terrain. *Nature*, **408**, 161–166.

European Union (2000) Directive 2000/60/EC of the European Parliament and of the council of 23 October 2000 establishing a framework for Community action in the field of water policy. *Official Journal of the European Communities*, **L327**, 1–73.

Evans, I.S. (1970) Salt crystallization and rock weathering: A review. *Revue de Geomorphologie Dynamique*, **19**, 153–177.

Fairbanks, R.G. (1989) A 17,000-year glacio-eustatic sea level record: Influence of glacial melting rates on the Younger Dryas event and deep-ocean circulation. *Nature*, **342**, 637–642.

Falco, R.E. (1977) Coherent motions in the outer region of turbulent boundary layers. *Physics of Fluids*, **20**, s124–s132.

Farrell, W.E. and Clark, J.A. (1976) On postglacial sea level. *Geophysical Journal of the Royal Astronomical Society*, **46**, 647–667.

Ferguson, R.I., Kirkbride, A.D. and Roy, A.G. (1996) Markov analysis of velocity fluctuations in gravel-bed rivers. In P.J. Ashworth, S.J. Bennett, J.L. Best and S.J. McLelland (eds), *Coherent flow structures in open channels*. Wiley, Chichester, 165–183.

Feyerband, P. (1975) How to defend society against science. *Radical Philosophy*, **11**, 3–8.

Feyerband, P. (1978) *Science in a free society*. New Left Books, London.

Firestone, R.B., West, A., Kennett, J.P., Becker, L., Bunch, T.E., Revay, Z.S., Schultz, P.H., Belgya, T., Kennett, D.J., Erlandson, J.M., Dickenson, O.J., Goodyear, A.C., Harris, R.S., Howards, G.A., Kloosterman, J.B., Lechler, P., Mayewski, P.A., Montgomery, J., Poreda, R., Darrah, T., Que Hee, S.S., Smith, A.R., Stich, A., Topping, W., Wittke, J.H. and Wolbach, W.S. (2007) Evidence for an extraterrestrial impact 12,900 years ago that contributed to megafaunal extinctions and the Younger Dryas cooling. *Proceedings of the National Academy of Sciences*, **104**, 16016–16021.

Fleming, K., Johnston, P., Zwartz, D., Yokoyama, Y., Lambeck, K. and Chappell, J. (1998) Refining the eustatic sea-level curve since the Last Glacial Maximum using far- and intermediate-field sites. *Earth and Planetary Science Letters*, **163**, 327–342.

Foley, R. (1999) Pattern and process in hominid evolution. In J. Bintliff (ed.), *Structure and contingency: Evolutionary processes in life and human society*. Leicester University Press, London, 31–42.

Frodeman, R.L. (1995) Geological reasoning: Geology as an interpretative and historical science. *Bulletin Geological Society of America*, **107**, 960–968.

Frodeman, R.L. (1996) Envisioning the outcrop. *Journal of Geoscience Education*, **44**, 417–427.

Fryirs, K. (2012) (Dis)connectivity in catchment sediment cascades: A fresh look at the sediment delivery problem. *Earth Surface Processes and Landforms*, DOI: 10.1002/esp.3242.

Gehrels, W.R., Belknap, D.F., Pearce, B.R. and Gong, B. (1995) Modeling the contribution of M_2 tidal amplification to the Holocene rise of mean high water in the Gulf of Maine and the Bay of Fundy. *Marine Geology*, **124**, 71–85.

Gehrels, W.R., Dawson, D.A., Shaw, J. and Marshall, W.A. (2011) Using Holocene relative sea-level data to inform future sea-level predictions: An example from southwest England. *Global and Planetary Change*, **78**, 116–126.

Giere, R.N. (1988) *Explaining science: A cognitive approach*. University of Chicago Press, Chicago.

Gilbert, G.K. (1877) *Report on the geology of the Henry Mountains*. Published in history of geology collection. 1978. Arno Press, New York.

Gilbert, G.K. (1896) The origin of hypotheses, illustrated by the discussion of a topographical problem. *Science*, **3**, 1–13.

Gill, J.L., Blois, J.L., Goring, S., Marlon, J.R., Bartlein, P.J., Nicoll, K., Scott, A.C. and Whitlock, C. (2012) Paleoecological changes at Lake Cuizeo were not consistent with an extraterrestrial impact. *Proceedings of the Natural Academy of Sciences*, **109**, E2243.

Glass, D.H. (2007) Coherence measures and inference to the best explanation. *Synthese*, **157**, 275–296.

Glass, D.H. (2012) Inference to the best explanation: Does it track truth? *Synthese*, **185**, 411–427.

Gleick, J. (1987) *Chaos*. Minerva, London.

Goodman, N. (1958) The test of simplicity. *Science*, **128**, 1064–1068.

Goodman, N. (1960) The way the world is. *Review of Metaphysics*, **14**, 48–56.

Goodman, N. (1967) Uniformity and simplicity. In *Uniformity and simplicity, Special Paper, Geological Society of America*, **89**, 93–99.

Goodman, N. (1978) *Ways of worldmaking*. Harvester Press, Brighton.

Goodstein, D. (2010) *On fact and fraud: Cautionary tales from the front lines of science*. Princeton University Press, Princeton.

Goudie, A. (1974) Further experimental investigation of rock weathering by salt and other mechanical processes. *Zeitschrift fur Geomorphologie Supplement*, **21**, 1–12.

Gould, S.J. (1965). Is uniformitarianism necessary? *American Journal of Science*, **263**, 223–228.

Gould, S.J. (1977) *Ontogeny and phylogeny*. The Belknap Press of Harvard University Press, Cambridge, Massachusetts.

Gould, S.J. (1987) *Time's arrow, time's cycle*. Penguin, London.

Gould, S.J. (2000) An awful terrible dinosaurian irony. In S.J. Gould *The lying stones of Marrakech*. Jonathan Cape, London, 183–200.

Gould, S.J. (2001). *Rock of ages*. Jonathan Cape, London.

Gregory, D. (1978) *Ideology, science and human geography*. Hutchinson, London.

Gregory, K. (1985). *The nature of physical geography*. Edward Arnold, London.

Guehlstorf, N. and Hallstrom, L.K. (2011) Participatory watershed management: A case study from maritime Canada. *Environmental Practice*, **14**, 143–153.

Gupta, A. and Ferguson, J. (eds) (1997) *Anthropological locations: Boundaries and frontiers of a field science*. University of California Press, Berkeley.

Hack, J.T. (1960) Interpretation of erosional topography in humid temperate regions. *American Journal of Science*, **258-A**, 80–97.

Hacking, I. (1983) *Representing and intervening*. Cambridge University Press, Cambridge.

Hacking, I. (1991) A tradition of natural kinds. *Philosophical Studies*, **61**, 109–126.

Hacking, I. (1999) *The social construction of what?* Harvard University Press, Cambridge, Massachusetts.

Haigh, M.J. (1987) The holon: Hierarchy theory and landscape research. *Catena Supplement*, **10**, 181–192.

Haines-Young, R. and Petch, J. (1986) *Physical Geography: Its nature and methods*. Harper and Row, London.

Haines-Young, R.H. and Potschin, M. (2010) The links between biodiversity, ecosystem services and human well-being. In D. Raffaelli and C. Frid (eds), *Ecosystem ecology: New synthesis*. BES Ecological Reviews Series. Cambridge University Press, Cambridge, 110–139.

Hall, K. (2004) Evidence for freeze-thaw events and their implications for rock weathering in northern Canada. *Earth Surface Processes and Landforms*, **29**, 43–57.

Hall, A.D. and Fagan, R.E. (1956) Definition of system. *General Systems Yearbook,* **1**, 18–28.

Hanebuth, T., Stattegger, K. and Grootes, P.M. (2000) Rapid flooding of the Sunda Shelf: A late-glacial sea-level record. *Science*, **288**, 1033–1035.

Hardiman, M., Scott, A.C., Collinson, M.E. and Anderson, R.S. (2012) Inconsistent redefining of the carbon spherule 'impact' proxy. *Proceedings of the National Academy of Sciences*, **109**, E2244.

Hardin, G. (1960) The competitive exclusion principle. *Science*, **131**, 1292–1297.

Harding, S. (1986) *The Science Question in Feminism*. Cornell University Press, Ithaca.

Harman, G. (1965) The inference to the best explanation. *The Philosophical Review*, **74**, 88–95.

Harre, R. (1985) *The philosophy of science*. Second edition. Oxford University Press, Oxford.

Harre, E.H. and Madden, R. (1973) In defence of natural agents. *Philosophical Quarterly*, **23**, 257–270.

Harrison, S. (2001) On reductionism and emergence in geomorphology. *Transactions of the Institute of British Geographers*, **26**, 327–339.

Harrison, J.L. (2011) Parsing 'participation' in action research: Navigating the challenges of lay involvement in technically complex participatory science projects. *Society and Natural Resources*, **24**, 702–716.

Harrison, S. and Dunham, P. (1998) Decoherence, quantum theory and their implications for the philosophy of geomorphology. *Transactions of the Institute of British Geographers*, **23**, 501–524.

Hartshorne, R. (1959) *Perspectives on the nature of geography*. McNally and Company, Chicago.

Harvey, D. (1969) *Explanation in geography*. Edward Arnold, London.

Harvey, D. (1994) The social construction of space and time: A relational theory. *Geographical Review of Japan, Series B*, **67**, 126–135.

Harvey, D. (2010) *The enigma of capital*. Profile Books, New York.

Hempel, C. (1965) *Aspects of scientific explanation and other essays in the philosophy of science*. Free Press, New York.

Herbertson, A.J. (1905). The major natural regions: An essay in systematic geography. *Geographical Journal*, **25**, 300–301.

Hesse, M. (1966) *Models and analogues in science*. Notre Dame University Press, New York.

Hester, A.J., Miles, J. and Gimingham, C.H. (1991a) Succession from heather moorland to birch woodland. I. Experimental alteration of specific environmental conditions in the field. *Journal of Ecology*, **79**, 303–315.

Hester, A.J., Miles, J. and Gimingham, C.H. (1991b) Succession from heather moorland to birch woodland. II. Growth and competition between *Vaccinium myrtillus, Deschampsia flexuosa* and *Agrostis capillaris. Journal of Ecology*, **79**, 317–327.

Hester, A.J., Gimingham, C.H. and Miles, J. (1991c). Succession from heather moorland to birch woodland. III. Seed availability, germination and early growth. *Journal of Ecology*, **79**, 329–344.

Hilderband, A.R., Penfield, G.T., Kring, D.A., Pilkington, M., Camargo, A., Jacobsen, S.B. and Boynton, W.V. (1991) Chicxulub crater: A possible Cretaceous/Tertiary boundary impact crater on the Yucatán Peninsula, Mexico. *Geology*, **19**, 867–871.

Holland, P.W. (1986) Statistics and causal inference. *Journal of the America Statistical Association*, **81**, 945–960.

Hooper, J. (2002) *Of moths and men: An evolutionary tale*. W.W. Norton and Co., New York.

Howard, A.D. (1988) Equilibrium models in geomorphology. In M.G. Anderson (ed.), *Modelling geomorphological systems*. John Wiley and Sons, Chichester, 49–72.

Huggett, R. (1980) *Systems analysis in geography*. Clarendon Press, Oxford.

Huisman, J. and Weissing, F.J. (1999) Biodiversity of plankton by species oscillations and chaos. *Nature*, **402**, 407–410.

Hull, D. (1976) A matter of individuality. *Philosophy of Science*, **45**, 335–360.

Hull, D. (1981) Units of evolution: A metaphysical essay. In R. Jensen and R. Harre (eds), *The philosophy of evolution*. Harvester, Brighton, 23–44.

Huntley, B. and Birks, H.J.B. (1983) *An atlas of past and present pollen maps of Europe 0-13,000 years ago*. Cambridge University Press, London.

Hutchinson, G.E. (1961) The paradox of the plankton. *The American Naturalist*, **95**, 137–145.

Huxley, T.H. (1877). *Physiography: an introduction to the study of nature*. Macmillan, London.

Hyatt, A. (1897) Cycle in the life of an individual (ontogeny) and in the evolution of its own group (phylogeny). *Proceedings American Academy of Arts and Science*, **32**, 209–224.

Hyndman, J. (2001) The field as here and now, not there and then. *Geographical Review*, **91**, 262–272.

Indermuhle, A., Stocker, T., Joos, F., Fischer, H., Smith, H.J., Wahlen, M., Deck, B., Masttroianni, D., Blunier, T., Meyer, R. and Stauffer, B. (1999) Holocene carbon-cycle dynamics based on CO_2 trapped in ice at Taylor Dome, Antarctica. *Nature*, **398**, 121–126.

Inkpen, R. (2007) Interpretation of erosion on rock surfaces. *Area*, **39**, 31–42.

Inkpen, R. and Petley, D. (2001) Fitness spaces and their potential for visualizing change in the physical landscape. *Area*, **33**, 242–251.

Inkpen, R. and Wilson, G.P. (2009) Explaining the past: Abductive and Bayesian reasoning. *The Holocene*, **19**, 329–334.

Inkpen, R.J., Collier, P. and Fontana, D. (2000) Close-range photogrammetric analysis of rock surfaces. *Zeitschrift fur Geomorphologie* **Suppl. Bd. 120**, 67–81.

Inkpen, R.J., Petley, D. and Murphy, W. (2004) Durability and rock properties. In B.J. Smith and A.V. Turkington (eds), *Stone decay: Its causes and controls*. Donhead, London.

International Council for Science (ICSU) (2002) *Science and technology for sustainable development*. World summit on Sustainable Development Volume 9.

Israde-Alcántara, I., Bischoff, J.L., Dominguez-Vázquez, G., Li, H.-C., DeCarli, P.S., Bunch, T.E., Wittke, J.H., Weaver, J.C., Firestone, R.B., West, A., Kennett, J.P., Mercer, C., Xie, S., Richman, E.K., Kinzie, C.R. and Wolbach, W.S. (2012a) Evidence from central Mexico supporting the Younger Dryas extraterrestrial impact hypothesis. *Proceedings of the National Academy of Sciences*, **109**, E738–E747.

Israde-Alcántara, I., Bischoff, J.L., Dominguez-Vázquez, G., Li, H.-C., DeCarli, P.S., Dominguez-Vázquez, G., Bunch, T.E., Firestone, R.B., Kennett, J.P. and West, A. (2012b) Reply to Blaauw *et al.*, Boslough, Daulton, Gill *et al.*, and Hardiman *et al.*: Younger Dryas impact proxies in Lake Cuitzeo, Mexico. *Proceedings of the National Academy of Sciences*, **109**, E2245–E2247.

ISRIC (1994) *Soil map of the world. Revised legend with corrections.* ISRIC, Wageningen.

Jacobsen, R.B., Wilson, D.C.K. and Ramirez-Monsalve, P. (2012) Empowerment and regulation – dilemmas in participatory fisheries science. *Fish and Fisheries*, **13**, 291–302.

Jeffreys, H. (1952) *The Earth: Its origin, history and physical constitution.* Cambridge University Press, Cambridge.

Jeffreys, H. (1970) Imperfections of elasticity and continental drift. *Nature*, **225**, 1007–1008.

Johnson, D. (1933) Role of analysis in scientific investigation. *Bulletin of the Geological Society*, **64**, 461–494.

Johnston, R.J. (1986) *Geography and geographers.* Arnold, London.

Joly, J. (1902) Some experiments on denudation by solution in fresh and salt water. *Proceedings of the Irish Academy. Section A: Mathematical and Physical Sciences*, **24**, 21–33.

Jonsson, A.C., Anderson, L., Olsson, J.A. and Johansson, M. (2011) Defining goals in participatory water management: Merging local visions and expert judgments. *Journal of Environmental Planning and Management*, **54**, 909–935.

Joos, F., Gerber, S., Prentice, I.C., Otto-Bleisner, B.L. and Valdes, P. (2004) Transient simulations of Holocene atmospheric carbon dioxide and terrestrial carbon since the Last Glacial Maximum. *Global Biogeochemical Cycles*, **18**, GB2002.

Kates, R.W., Clark, W.C., Corell, R., Hall, J.M., Jaeger, C.C., Lowe, I., McCarthy, J.J., Schellnhuber, H.J., Bolin, B., Dickson, N.M., Faucheux, S., Gallopin, G.C., Grubler, A., Huntley, B., Jager, J., Jodha, N.S., Kasperson, R.E., Mabogunje, A., Matson, P., Mooney, H., Moore, B., O'Riordan, T. and Svedin, U. (2001) Environment and development: Sustainability science. *Science*, **292**, 641–642.

Kauffman, S.A. (2000) *Investigations.* Oxford University Press, Oxford.

Kelly, M., Bennion, H., Burgess, A., Ellis, J., Juggins, S., Guthrie, R., Jamieson, J., Adriaenssens, V. and Yallop, M. (2009). Uncertainty in ecological status assessments of lakes and rivers using diatoms. *Hydrobiologia*, **633**, 5–15.

Kennedy, B.A. (1992). Hutton to Horton: views of sequence, progression and equilibrium in geomorphology. *Geomorphology*, **5**, 231–250.

Kettlewell, H.B.D. (1952) Use of radioactive traces in the study of insect populations (*Lepidoptera*). *Nature*, **170**, 584–585.

Kettlewell, H.B.D. (1955) Selection experiments on industrial melanism in the *Lepidoptera*. *Heredity*, **9**, 323–342.

Kettlewell, H.B.D. (1956) Further selection experiments on industrial melansim in the *Lepidotera. Heredity*, **10**, 287–301.

Kettlewell, H.B.D. (1973) *The evolution of melanism: The study of a recurring necessity.* Clarendon Press, Oxford.

Kirkbride, A.D. and Ferguson, R.I. (1995) Turbulent flow structure in a gravel-bed river. Markov chain analysis of the fluctuating velocity profile. *Earth Surface Processes and Landforms*, **20**, 721–733.

Kobayashi, A. and Mackenzie, S. (1989) *Remaking human geography.* Unwin Hyman, Boston.

Kohler, R.E. (2002) *Landscapes and labscapes: Exploring the lab-field border in biology.* University of Chicago Press, Chicago.

Kohler, R.E. (2012) Practice and place: Twentieth century field biology. *Journal of the History of Biology*, **45**, 579–586.

Komiyama, H. and Takeuchi, K. (2006) Sustainability science: Building a new discipline. *Sustainability Science*, **1**, 1–6.

Koppen, W. (1900) Versuch einer Klassifikation der Klimate, vorzugsweise nach ihren Beziehungen zur Pflanzenwelt. *Geographische Zeitschrift*, **6**, 593–611.

Kripe, S. (1980) *Naming and necessity*. Harvard University Press, Cambridge, Mass.

Kuchar, J., Milne, G., Hubbard, A., Patton, H., Bradley, S., Shennan, I. and Edwards, R. (2012) Evaluation of a numerical model of the British–Irish ice sheet using relative sea-level data: Implications for the interpretation of trimline observations. *Journal of Quaternary Science*, **27**, 597–605.

Kuchler, A.W. (1954). Plant geography. In *American Geography: inventory and prospect*. James, P.E. and Jones, C.F. (eds.). Syracuse University Press, Syracuse, NY. 429–440.

Kuhn, T.S. (1962) *The structure of scientific revolutions*. University of Chicago Press, Chicago.

Kuhn, T.S. (1977) *The essential tension*. University of Chicago Press, Chicago.

Lahr, J. and Kooistra, L. (2010) Environmental risk mapping of pollutants: State of the art and communication aspects. *Science of the Total Environment*, **408**, 3899–3907.

Lakatos, I. (1970) Falsification and the methodology of scientific research programmes. In I. Lakatos and A. Musgrave (eds), *Criticism and the growth of knowledge*. Cambridge University Press, Cambridge, 91–196.

Lamarque, F. Ouetier, F. and Lavorel, S. (2011) The diversity of the ecosystem services concept and its implications for their assessment and management. *Comptes Rendues Biologies*, **334**, 441–449.

Lambeck, K. (1991) Glacial rebound and sea-level change in the British Isles. *Terra Nova*, **3**, 379–389.

Lambeck, K. (1993) Glacial rebound of the British Isles II. A high-resolution, high-precision model. *Geophysical Journal International*, **115**, 960–990.

Lambeck, K. (1995) Late Devensian and Holocene shorelines of the British Isles and North Sea from models of glacio-hydro-isostatic rebound. *Journal of the Geological Society of London*, **152**, 437–448.

Lane, S.N. (2001) Constructive comments on D. Massey 'Space-time, "science" and the relationship between physical geography and human geography'. *Transactions of the Institute of British Geographers*, **26**, 243–256.

Lane, S.N. and Richards, K.S. (1997) Linking river channel form and process: Time, space and causality revisited. *Earth Surface Processes and Landforms*, **22**, 249–260.

Lane, S.N., Bradbrook, K.F., Richards, K.S., Biron, P.A. and Roy, A.G. (1999) The application of computational fluid dynamics to natural river channels: Three-dimensional versus two-dimensional approaches. *Geomorphology*, **29**, 1–20.

Lane, S.N., Odoni, N. Landstrom, C., Whatmore, S.J., Ward, N. and Bradley, S. (2010) Doing flood risk science differently: An experiment in radical scientific method. *Transactions of the Institute of British Geographers*, **36**, 15–36.

Laoire, C.N. and Shelton, N.J. (2003) 'Contracted out': Some implications of the casualization of academic labour in geography. *Area*, **35**, 92–100.

Latour, B. (2005) *Reassembling the social – an introduction to actor-network theory*. Oxford University Press, Oxford.

Lawson, I.T., Frogley, M.R., Bryant, C., Preece, R. and Tzedakis, P. (2004) The Lateglacial and Holocene environmental history of the Ioannina basin, north-west Greece. *Quaternary Science Reviews*, **23**, 1599–1625.

Lawson, I.T., Gathorne-Hardy, F.J., Church, M.J., Newton, A.J., Edwards, K.J., Dugmore, A.J. and Einarsson, A. (2007) Environmental impacts of the Norse settlement: Palaeoenvironmental data from Mývatnssveit, northern Iceland. *Boreas*, **36**, 1–19.

Leira, M., Jordan, P., Taylor, D., Dalton, C., Bennion, H., Rose, N. and Irvine, K. (2006) Assessing the ecological status of candidate reference lakes in Ireland using palaeolimnology. *Journal of Applied Ecology*, **43**, 816–827.

Lengwiler, M. (2008) Participatory approaches in science and technology: Historical origins and current practices in critical perspective. *Science Technology and Human Values*, **33**, 186–200.

Leopold, L.B. and Langbein, W.B. (1963) Association and indeterminacy in geomorphology. In C.C. Albritton (ed.), *The fabric of geology*. Addison-Wesley, Reading, Mass., 184–192.

Leopold, L.B., Wolman, M.G. and Miller, P. (1964). *Fluvial processes in geomorphology*. WH Freeman, San Francisco.

Lepistö, L., Holopainen, A-L., Vuoristo, H. and Rekolainen, S. (2006) Phytoplankton assemblages as a criterion in the ecological classification of lakes in Finland. *Boreal Environmental Research*, **11**, 35–44.

Levins, R. (1966) The strategy of model building in population biology. In E. Sober (ed.), *Conceptual issues in evolutionary biology*. MIT Press, Cambridge, MA, 18–27.

Levins, R. and Lewontin, R. (1985) *The dialectical biologist*. Harvard University Press, Harvard.

Lewis, D. (1979) Counterfactual dependence and time's arrow. *Nous*, **13**, 455–476.

Lewontin, R.C. (1995) Genes, environment and organisms. In Silvers, R.B. (ed.), *Hidden histories of science*. Granta Books, London, 115–140.

Li, C. (1993) Natural kinds: Direct reference, realism and the impossibility of necessary *a posteriori* truth. *Review of Metaphysics*, **47**, 262–276.

Linton, D.L. (1955) The problem of tors. *Geographical Journal*, **121**, 470–487.

Lipton, P. (2004) *Inference to best explanation*. Second edition. Routledge, London.

Livingstone, D.N. (1984) Natural theology and neo-Lamarckianism: The changing context of nineteenth century geography in the United States and Great Britain. *Annals of the Association of American Geographers*, **74**, 9–28.

Long, A.J., Roberts, D.H. and Rasch, M. (2003) New observations on the relative sea level and deglacial history of Greenland from Innaarsuit, Disko Bugt. *Quaternary Research*, **60**, 162–171.

Long, A.J., Woodroffe, S.A., Milne, G.A., Bryant, C.L. and Wake, L.M. (2010) Relative sea level change in west Greenland during the last millennium. *Quaternary Science Reviews*, **29**, 367–383.

Lorenz, E.N. (1963) Deterministic nonperiodic flow. *Journal of Atmospheric Sciences*, **20**, 130–141.

Lorenz, E.N. (1982) Atmospheric predictability experiments with a large numerical model. *Tellus*, **34**, 505–513.

Loux, M.J. (1998) *Metaphysics*. Routledge, London.

Lyell, C. (1833) *The principles of geology*. John Murray, Edinburgh.

Mackin, J.H. (1963) Rational and empirical methods of investigation in geology. In C.C. Albritton (ed.), *The fabric of geology*. Addison-Wesley, Reading, Mass., 135–163.

Mackinder, H.J. (1887) On the scope and methods of geography. *Proceedings of the Royal Geographical Society*, **9**, 141–160.

Maisonneuve, N., Stevens, M. and Ochab, B. (2010) Participatory noise pollution monitoring using mobile phones. *Information Polity*, **15**, 51–71.

Majerus, M.E.N. (1998) *Melanism: Evolution in action*. Oxford University Press, Oxford.

Manson, S.M. (2001) Simplifying complexity: A review of complexity theory. *Geoforum*, **32**, 405–414.

Massey, D. (1999) Space-time, 'science' and the relationship between physical geography and human geography. *Transactions of the Institute of British Geographers*, **24**, 261–276.

Masterman, M. (1970) The nature of a paradigm. In I. Lakatos and A. Musgrave (eds), *Criticism and the growth of knowledge*. Cambridge University Press, Cambridge, 59–90.

May, R.M. (1974) Biological populations with nonoverlapping generations: Stable points, stable cycles, and chaos. *Science*, **186**, 645–647.

May, R.M. (1976) Simple mathematical models with very complicated dynamics. *Nature*, **261**, 459–467.

Mayer, L. (1992) Some comments on equilibrium concepts and geomorphic systems. *Geomorphology*, **5**, 277–295.

McGreevy, J.P. (1982) 'Frost and salt' weathering: Further experimental results. *Earth Surface Processes and Landforms*, **7**, 475–488.

Melton, M.A. (1958) Correlation structure of morphometric properties of drainage systems and their controlling agents. *Journal of Geology*, **66**, 442–460.

Merchant, C. (1980) *The death of nature: Women, ecology and scientific revolution.* HarperCollins, New York.

Merchant, C., in press. Francis Bacon and the 'vexations of art': Experimentation as intervention. *The British Journal for the History of Science*.

Merrill, G.P. (1896) The principles of rock weathering. *The Journal of Geology*, **4**, 850–871.

Metcalfe, S.E., Whyatt, J.D., and Derwent, R.G. (1995) A comparison of model and observed network estimates of sulphur deposition across Great Britain for 1990 and its likely source attribution. *Quarterly Journal of the Royal Meteorological Society*, **121**, 1387–1411.

Metcalfe, S.E., Whyatt, J.D., Broughton, R., Derwent, R.G., Finnegan, D., Hall, J., Mineter, M., O'Donoghue, M. and Sutton, M.A. (2001) Developing the Hull Acid Rain Model: Its validation and implications for policy makers. *Environmental Science and Policy*, **4**, 25–37.

Middelburg, J.J. and Nieuwenhuize, J. (1998) Carbon and nitrogen stable isotopes in suspended matter and sediments from the Schelde Estuary. *Marine Chemistry*, **60**, 217–225.

Millennium Ecosystem Assessment (MA) (2005) *Ecosystems and human well-being: Synthesis*. Island Press, Washington D.C.

Minkler, M. (2010) Linking science and policy through community-based participatory research to study and address health disparities. *American Journal of Public Health*, **100**, S81–S87.

Mitrovica, J.X. and Peltier, W.R. (1991) On postglacial geoid subsidence over the equatorial oceans. *Journal of Geophysical Research*, **96**, 20053–20071.

Mitrovica, J.X. and Milne, G.A. (2002) On the origin of late Holocene sea-level highstands within equatorial ocean basins. *Quaternary Science Reviews*, **21**, 2179–2190.

Mitrovica, J.X. and Milne, G.A. (2003) On post-glacial sea level: I. General theory. *Geophysical Journal International*, **154**, 253–267.

Moh'd, B.K., Howarth, R.J. and Bland, C.H. (1996) Rapid prediction of Building Research Establishment limestone durability class from porosity and saturation. *Quarterly Journal of Engineering Geology*, **29**, 285–297.

Murdoch, J. (1998) The spaces of actor-network theory. *Geoforum*, **29**, 357–374.

Murray, A.B., Lazarus, E., Ashton, A., Baas, A., Cuco, G., Coulthard, T., Fonstad, M., Haff, P., McNamara, D., Paola, C., Pelletier, J. and Reinhardt, L. (2009) Geomorphology, complexity, and the emerging science of the Earth surface. *Geomorphology*, **103**, 496–505.

Murton, J.B., Bateman, M.D., Dallimore, S.R., Teller, J.T. and Yang Z. (2010) Identification of Younger Dryas outburst flood path from Lake Agassiz to the Arctic Ocean. *Nature*, **464**, 740–743.

Nakagawa, H. and Nezu, I. (1981) Structures of space-time correlations of bursting phenomena in an open channel flow. *Journal of Fluid Mechanics*, **104**, 1–43.

Newman, M.E.J. and Watts, D.J. (1999a) 'Renormalization group analysis of the small-world network model', *Physics Letters A*, **263**, 341–346.

Newman, M.E.J. and Watts, D.J. (1999b) 'Scaling and percolation in the small-world network model', *Physical Review E*, **60**, 7332–7342.

Nichols, D.J., Jarzen, D.M., Orth, C.J. and Oliver, P.O. (1986) Palynological and iridium anomalies at Cretaceous-Tertiary boundary, south-central Saskatchewan. *Science*, **231**, 714–717.

Oldfield, F. (1993) Forward to the past: Changing approaches to Quaternary palaeoecology. In F.M. Chambers (ed.), *Climate change and human impact on the landscape*. Chapman and Hall, London, 13–21.

Oldroyd, D. (1986) *The arch of knowledge: The history of the philosophy and methodology of science*. Methuen, New York.

Olsson, E.J. (2002) What is the problem with coherence and truth? *Journal of Philosophy*, **99**, 246–272.

O'Neill, R.V., DeAngelis, D.L., Waide, J.B. and Allen, T.F.H. (1980) *A hierarchical concept of ecosystems*. Princeton University Press, Princeton, NJ.

Oreskes, N. (ed.) (2003) *Plate tectonics: An insider's history of the modern theory of the Earth*. Westview Press, Boulder, California.

Orzack, S.H. and Sober, E. (1993) A critical assessment of Levin's 'The strategy of model building in population biology' (1966). *Quarterly Review of Biology*, **68**, 533–546.

Outram, D. (1996) New spaces in natural history. In N. Jardine, J.A. Secord and E.C. Spary (eds), *Cultures of natural history*. Cambridge University Press, Cambridge, 249–265.

Outram, D. (1999) On being Perseus: New knowledge, dislocation, and Enlightenment exploration. In D.N. Livingstone and C.W.J. Withers (eds), *Geography and Enlightenment*. Chicago University Press, Chicago, 281–294.

Page, T., Whyatt, J.D., Beven, K.J. and Metcalfe, S.E. (2004) Uncertainty in modeled estimates of acid deposition across Wales: A GLUE approach. *Atmospheric Environment*, **38**, 2079–2090.

Paola, C. (2001) Modelling stream braiding over a range of scales. In M.P. Mosely (ed.), *Gravel-bed rivers V*, New Zealand Hydrological Society, Wellington, New Zealand. 11–38.

Pawson, R. (1989) *Measure for measures*. Routledge, London.

Pearl, J. (2000) *Causality*. Cambridge University Press, Cambridge.

Peltier, L.C. 1950. The Geographic Cycle in Periglacial Regions as it is Related to Climatic Geomorphology. Annals of the Association of American Geographers, 49, 214-236.

Peltier, W.R. (1998) Postglacial variations in the level of the sea: Implications for climate dynamics and solid earth geophysics. *Reviews of Geophysics*, **36**, 603–689.

Peltier, W.R. (2002) Global glacial isostatic adjustment: Palaeogeodetic and space-geodetic tests of the ICE-4G (VM2) model. *Journal of Quaternary Science*, **17**, 491–510.

Peltier, W.R. (2004) Global glacial isostasy and the surface of the ice-age Earth: The ICE-6G (VM2) Model and GRACE. *Annual Review of Earth and Planetary Sciences*, **32**, 111–149.

Peltier, W.R., Shennan, I., Drummond, R. and Horton, B. (2002) On the postglacial isostatic adjustment of the British Isles and the shallow viscoelastic structure of the Earth. *Geophysical Journal International*, **148**, 443–475.

Penck, W. (1924) *Die morphologishce Abalyse. Ein Kapital der physikalischen Geologie*. Engelhorns, Stuttgart. English translation by H. Czech and K.C. Boswell 1953. *Morphological analysis of landforms*. Macmillan, London.

Peters, R.H. (1991) *A critique for ecology*. Cambridge University Press, Cambridge.

Phillips, J.D. (1986) Sediment storage, sediment yield, and time scales in landscape denudation studies. *Geographical Analysis*, **18**, 161–167.

Phillips, J.D. (1988) The role of spatial scale in geomorphic systems. *Geographical Analysis*, **20**, 308–317.

Phillips, J. (1995) Nonlinear dynamics and the evolution of relief. *Geomorphology*, **14**, 57–64.

Phillips, J.D. (1999) Divergence, convergence, and self-organization in landscapes. *Annals of the Association of American geographers*, **89**, 466–488.

Phillips, J.D. (2001) Methodology, scale and the field of dreams. *Annals of the Association of American Geographers*, **91**, 754–760.

Phillips, J.D. (2003) Sources of nonlinearity and complexity in geomorphic systems. *Progress in Physical Geography*, **27**, 1–23.

Phillips, J.D. (2011) Emergence and pseudo-equilibrium in geomorphology. *Geomorphology*, **132**, 319–326.

Pierce, C.S. (1957) *Essays in the philosophy of science*. Bobbs-Merrill, New York.

Pile, A., Heinonen, P., Karttunen, K., Koskenniemi, E., Lepistö, L., Pietiläinen, O.-P., Rissanen, J. and Vuoristo, H. (2002) Finnish draft for typology of lakes and rivers. *TemaNord*, **566**, 42–43.

Pilke, A., Heinonen, P., Karttunen, K., Koskenniemi, E., Lepistö, L., Pietiläinen, O.-P., Rissanen, J. and Vuoristo, H. (2002). Finnish draft for typology of lakes and rivers. *TemaNord*, **566**, 42–43.

Pitman, A.J. (2005) On the role of geography in Earth Science Systems. *Geoforum*, **36**, 137–148.

Platt, J. (1970) Hierarchical restructuring. *General Systems*, **15**, 49–54.

Popper, K. (1963) *Conjectures and refutations: The growth of scientific knowledge*. Routledge, London.

Popper, K.R. (1968) *The logic of scientific discovery*. Hutchinson, London.

Potschin, M.B. and Haines-Young, R.H. (2011) Ecosystem services: Exploring a geographical perspective. *Progress in Physical Geography*, **35**, 575–594.

Powell, R.C. (2002) The Sirens' voices: Field practices and dialogue in geography. *Area*, **34**, 261–272.

Price, C.A., Weitz, J.S., Savage, V.M., Stegen, J., Clarke, A., Coomes, D.A., Dodds, P.S., Etienne, R.S., Kerkhoff, A.J., McCulloh, K., Niklas, K.J., Olff, H. and Swenson, N.G. (2012) Testing the metabolic theory of ecology. *Ecology Letters*, **15**, 1465–1474.

Putnam, H. (1973) The meaning of meaning. *Journal of Philosophy*, **70**, 699–711.

Putnam, H. (1994) Sense, nonsense and the senses: An inquiry into the powers of the human mind. *The Journal of Philosophy*, **91**, 445–471.

Raper, J. (2000) *Multidimensional geographic information science*. Taylor and Francis, London.

Raper, J. and Livingstone, D. (1995) Development of a geomorphological spatial model using object-oriented design. *International Journal of Geographical Information Systems*, **9**, 359–383.

Raper, J. and Livingstone, D. (2001) Let's get real: Spatio-temporal identity and geographic entities. *Transactions of the Institute of British Geographers*, **26**, 237–242.

Rapp, A. (1960). Recent development of mountain slopes in Karkevagge and surroundings, Northern Scandinavia. *Geografriska Annaler*, **42**, 65–200.

Refsgaard, J.C., Henriksen, H.J., Harrar, W.G., Scholten, H. and Kassahun, A. (2005) Quality assurance in model based water management – review of existing practice and outline of new approaches. *Environmental Modelling and Software*, **20**, 1201–1215.

Refsgaard, J.C., van der Sluijs, J.P., Hojberg, A.L. and Vanrolleghem, P.A. (2007) Uncertainty in the environmental modelling process – a framework and guidance. *Environmental Modelling and Software*, **22**, 1543–1556.

Reitsma, F. (2003) A response to simplifying complexity. *Geoforum*, **34**, 13–16.

Renssen, H., van Geel, B., van der Plicht, J. and Magny, M. (2000) Reduced solar activity as a trigger for the start of the Younger Dryas? *Quaternary International*, **68–71**, 373–383.

Rescher, N. (1991) Conceptual idealism revisited. *Review of Metaphysics*, **44**, 495–523.

Rescher, N. (1992) The promise of process philosophy. In R.W. Burch and H.J. Saatkamp (eds), *Frontiers in American Philosophy*. Texas A & M University Press, College Station, TX.

Rescher, N. (2001) Functionalistic pragmatism. *The Philosophical Forum*, **32**, 191–205.

Rhoads, B.L. and Thorn, C.E. (1996a) Observation in geomorphology. In B.L. Rhoads and C.E. Thorn (eds), *The scientific nature of geomorphology*. Wiley, Chichester, 21–56.

Rhoads, B.L. and Thorn, C.E. (1996b) Toward a philosophy of geomorphology. In B.L. Rhoads and C.E. Thorn (eds), *The scientific nature of geomorphology*. Wiley, Chichester, 115–143.

Rhoads, B.L. and Thorn, C.E. (1996c). *The scientific nature of geomorphology*. Wiley, Chichester.

Richards, K. (1990) 'Real' geomorphology. *Earth Surface Processes and Landforms*, **15**, 195–197.

Richards, K. (1996) Samples and cases: Generalization and explanation in geomorphology. In B.L. Rhoads and C.E. Thorn (eds), *The scientific nature of geomorphology*. Wiley, Chichester, 171–190.

Richards, K.S., Brooks, S.M., Clifford, N.J., Harris, T. and Lane, S. (1997) Theory, measurement and testing in 'real' geomorphology and physical geography. In D.R. Stoddart (ed.), *Process and from in geomorphology*. Routledge, London, 265–292.

Rind, D., Peteet, D., Broecker, W., McIntyre, A. and Ruddiman, W. (1986) The impact of cold North Atlantic sea surface temperatures on climate: Implications for the Younger Dryas cooling (11–10 k). *Climate Dynamics*, **1**, 3–33.

Robben, ACGM and Sluka, J.A. (eds.) (2012) *Ethnographic fieldwork: An anthropological reader*. Second edition. John Wiley and Sons, Chichester.

Roberts, M.J., Scourse, J.D., Bennell, J.D., Huws, D.G., Jago, C.F. and Long, B.T. (2011) Late Devensian and Holocene relative sea-level change in North Wales, UK. *Journal of Quaternary Science*, **26**, 141–155.

Rose, G. (1992) Geography as a science of observation: The landscape, the gaze and masculinity. In F. Driver and G. Rose (eds), *Nature and science: Essay in the history of geographical knowledge*. Historical Geography Research Series 28, Institute of British Geographers, London, 8–18.

Rose, G. (1993) Progress in geography and gender. Or something else? *Progress in Human Geography*, **17**, 531–537.

Rose, G. (1997) Situating knowledges: Positionality, reflexivities and other tactics. *Progress in Human Geography*, **21**, 305–320.

Rosenthal, S.B. (1994) *Charles Peirce's pragmatic pluralism*. State University of New York Press, New York.

Ross, K.D. and Butlin, R.N. (1989) *Durability tests for building stone*. Building Research Establishment Report 141. Building Research Establishment, Garston, UK.

Ross, K.D. and Massey, S. (1990) A response to Sedman and Stanley analysis. *Stone Industries*, **25**, 30–33.

Rouse, J. (1987) *Knowledge and power*. Cornell University Press, Ithaca.

Rouse, J. (2008) Laboratory fictions. In M. Suarez (ed.), *Fiction in science: Philosophical essays on modeling and idealization*. Routledge, New York, 37–55.

Roy, A.G., Buffin-Belanger, T. and Deland, S. (1996) Scales of turbulent coherent flow structures in a gravel bed river. In P.J. Ashworth, S.J. Bennett, J.L. Best and S.J. McLelland (eds), *Coherent flow structures in open channels*, Wiley, Chichester, 147–164.

Ruddiman, W.F. (2003) The anthropogenic greenhouse era began thousands of years ago. *Climatic Change*, **61**, 261–293.

Ruddiman, W.F., Kutzbach, J.E. and Vavrus, S.J. (2011) Can natural or anthropogenic explanations of late-Holocene CO_2 and CH_4 increases be falsified? *Holocene*, 21, 865–887.

Rudge, D.W. (2005) Did Kettlewell commit fraud? Re-examining the evidence. *Public Understanding of Science*, **14**, 249–268.

Ruse, M. (1987) Biological species: Natural kinds, individuals, or what? *British Journal for the Philosophy of Science*, **38**, 225–242.

Sabbagh, K. (1999) *A rum affair: A true story of botanical fraud*. Da Capo Press, Cambridge, Mass.

Salles, J.M. (2011) Valuing biodiversity and ecosystem services: Why put economic values on nature? *Comptes Rendues Biologies*, **334**, 469–482.

Salmon, W.C. (1998) Causality and explanation. Oxford University Press, New York.

Sauer, C.O. (1956) The agency of man. In W.L. Thomas (ed.), *Man's role in changing the face of the earth*. University of Chicago Press, Chicago, 49–69.

Sayer, A. (1992) *Method in social science: A realist approach*. Second edition. Routledge, London.

Schaefer, F.K. (1953) Exceptionalism in geography: A methodological examination. *Annals of the Association of American Geographers*, **43**, 226–245.

Schaffer, W.M. (1981) Ecological abstractions: The consequences of reduced dimensionality in ecological models. *Ecological Monographs*, **5**, 383–401.

Schneider, S.H. (2001) A constructive deconstruction of the deconstructionists. *Annals of the Association of American Geographers*, **91**, 338–344.

Schumm, S.A. (1979) Geomorphic thresholds: The concept and its application. *Transactions of the Institute of British Geographers*, **4**, 485–515.

Schumm, S.A. (1991) *To interpret the Earth: Ten ways to be wrong*. Cambridge University Press, Cambridge.

Schumm, S.A. and Lichty, R.W. (1965) Time, space and causality in geomorphology. *American Journal of Science*, **263**, 110–119.

Schwartz, S.P. (1980) Natural kinds and nominal kinds. *Mind*, **89**, 182–195.

Sedman, J.H.F. and Stanley, L. (1990a) Variations in the physical properties of porous building limestones. *Stone Industries*, **25**, 22–24.

Sedman, J.H.F. and Stanley, L. (1990b) The crystallization test as a measure of durability. *Stone Industries*, **25**, 26–28.

Shain, R. (1993) Mill, Quine and natural kinds. *Metaphilosophy*, **24**, 275–292.

Shannon, C.E. and Weaver, W. (1949) *The mathematical theory of communication*. University of Illinois Press, Urbana.

Shennan, I. (1995) Sea-level and coastal evolution: Holocene analogues for future changes. *Coastal Zone Topics: Process, Ecology and Management*, **1**, 1–9.

Shennan, I., Hamilton, S., Hillier, C. and Woodroffe, S. (2005) A 16 000-year record of near-field relative sea-level changes, northwest Scotland, United Kingdom. *Quaternary International*, **133–134**, 95–106.

Shennan, I., Bradley, S., Milne, G., Brooks, A., Bassett, S. and Hamilton, S. (2006) Relative sea-level changes, glacial isostatic modelling and ice-sheet reconstructions from the British Isles since the Last Glacial Maximum. *Journal of Quaternary Science*, **21**, 585–599.

Shennan, I., Milne, G. and Bradley, S. (2009) Late Holocene relative land- and sea-level changes: Providing information for stakeholders. *GSA Today*, **19**, 52–53.

Simms, M.J. (2002) The origin of enigmatic, tubular, lake-shore Karren: A mechanism for rapid dissolution of limestone in carbonate-saturated waters. *Physical Geography*, **23**, 1–20.

Simpson, G.G. (1963) Historical science. In C.C. Albritton (ed.), *The fabric of geology*. Addison-Wesley, Reading, Mass., 24–48.

Smit, J. and Hertogen, J. (1980) An extra-terrestrial event at the Cretaceous-Tertiary boundary. *Nature*, **285**, 198–200.

Smol, J.P. and Stoermer, E.F. (2010) *The diatoms: Applications for the environmental and Earth sciences*. Cambridge University Press, Cambridge.

Smol, J.P., Wolfe, A.P., Birks, J.H.B., Douglas, M.S.V., Jones, V.J., Korhola, A., Pienitz, R., Rühland, K., Sorvari, S., Antoniades, D., Brooks, S.J., Fallu, M-A., Hughes, M., Keatley, B.E., laing, T.S., Michelutti, N., Nazarova, L., Nyman, M., Paterson, A.M., Perren, B., Quinlan, R., Rautio, M., Saulnier-Talbot, E., Siitonen, S., Solovieva, N. and Weckström, J. (2005) Climate-driven regime shifts in the biological communities of arctic lakes. *Proceedings of the National Academy of Sciences*, **102**, 4397–4402.

Sober, E. (1988) *Reconstructing the past: Parsimony, evolution and inference*. MIT Press, Cambridge, Mass.

Solomon, S. (2001). *The coldest March: Scott's fatal Antarctic expedition*. Melbourne University Press, Melbourne.

Spedding, N. (1999) On growth and form in geomorphology. *Earth Surface Processes and Landforms*, **22**, 261–265.

Stern, N. (2007) *The economics of climate change: The Stern review*. Cambridge University Press, Cambridge.

Stevens, P.A., Ormerod, S.J. and Reynolds, B. (1997) *Final report of the acid waters survey for Wales*. Institute of Terrestrial Ecology, Bangor, Wales.

Stocker, B., Strassman, K. and Joos, F. Sensitivity of Holocene atmospheric CO_2 and the modern carbon budget to early human land use: analyses with a process-based model. *Biogeosciences Discussions*, **7**, 921–952.

Stoddart, D.R. (1968). Climatic geomorphology: Review and assessment. *Progress in Geography*, **1**, 160–222.

Stoddart, D. R. (1975). That Victorian science': Huxley's *Physiography* and its impact on geography. *Transactions of the Institute of British Geographers*, **66**, 17–40.

Stoddart, D.R. (1981) The paradigm concept and the history of geography. In D.R. Stoddart (ed.), *Geography, ideology and social concern*. Blackwell, Oxford, 70–79.

Strahler, A.N. (1952) Dynamic basis of geomorphology. *Bulletin Geological Society of America*, **63**, 923–938.

Strahler, A.N. (1957) Quantitative analysis of watershed geomorphology. *Transactions of the American Geophysical Union*, **38**, 913–920.

Strahler, A.N. (1980) Systems theory in physical geography. *Physical Geography*, **1**, 1–27.

Suppe, F. (1977) The search for philosophical understanding of scientific theories. In Suppe, F. (ed.), *The structure of scientific theories*. University of Illinios Press, Illinois, 3–41.

Surovell, T.A., Holliday, V.T., Gingerich, J.A.M., Ketron, C., Vance Haynes Jr., C., Hilman, I., Wagner, D.P., Johnson, E. and Claeys, P. (2009) An independent evaluation of the Younger Dryas extraterrestrial impact hypothesis. *Proceedings of the National Academy of Sciences*, **106**, 18155–18158.

Swart, R.J., Raskin, P. and Robinson, J. (2004) The problem of the future: Sustainability science and scenario analysis. *Global Environmental Change*, **14**, 137–146.

Talent, J.A., (1989) The case of the peripatetic fossils. *Nature*, **338**, 613–615.

Tansley, A.G. (1935). The use and abuse of vegetational concepts and terms. *Ecology*, **16**, 284–307.

Teller, J. (1987) Lake Agassiz and its contribution to flow through the Ottawa-St. Lawrence System. *Geological Association of Canada Special Paper*, **35**, 281–289.

Terjung, W. (1976) Climatology for geographers. *Annals of the Association of American Geographers*, **66**, 199–222.

Thompson, P. (1983) The structure of evolutionary theory: A semantic approach. *Studies in the History and Philosophy of Science*, **14**, 215–219.

Thorn, C.E. (1988) *Introduction to theoretical geomorphology*. Unwin Hyman, London.

Thornwaite, C.W. (1948) An approach toward a rational classification of climate. *Geographical Review*, **38**, 55–94.

Tinkler, K.J. (1985) *A short history of geomorphology*. Croom Helm, London.

Tooth, S. (2006) Virtual globes: A catalyst for the re-enchantment of geomorphology? *Earth Surface Processes and Landforms*, **31**, 1192–1194.

Torok, A. and Prikryl, R. (2010) Current methods and future trends in testing, durability and provenance studies of natural stones used in historical monuments. *Engineering Geology*, **115**, 139–142.

Toulmin, S.E. (1960) *The philosophy of science*. Harper and Row, New York.

Townsend, W.N. (1973) *An introduction to the scientific study of the soil*. Fifth edition. Edward Arnold, London.

Tricart, J and Callieux, A. (1972). *An introduction to climatic geomorphology*. Longmans, London.

Trudgill, S.T. (1977). *Soil and Vegetation Systems*. Oxford University Press, Oxford.

Trudgill, S.T., Viles, H.A., Inkpen, R.J. and Cooke, R.U. (1989) Remeasurement of weathering rates, St Paul's Cathedral, London. *Earth Surface Processes and Landforms*, **14**, 175–196.

Trudgill, S.T., Viles, H.A., Cooke, R.U. ⌐
St Paul's Cathedral, London. *Atmospheric* ⌐

Trudgill, S.T., Viles, H.A., Inkpen, R.J., Moses, ⌐ 1990) Rates of stone loss at
D.I., Cooke, R.U. (2001) Twenty-year weather⌐ 361–363.
London. *Earth Surface Processes and Landforms,* ⌐ ates, T., Collier, P., Smith,

Turkington, A. (1996) Stone durability. In *Processes o⌐* ⌐ ts at St Paul's Cathedral,
Warke, P.A. (eds.). Donhead, London. 19–31.

Turner, J.R.G. (1985) Fisher's evolutionary faith and the cha⌐ ecay. Smith, B.J. and
and M. Ridley (eds), *Oxford surveys in evolutionary biolog⌐*
Press, Oxford, 159–195. ⌐ icry. In R. Dawkins

Turner, R.E. (1997) Wetland loss in the Northern Gulf of Mexico: ⌐ xford University
eses. *Estuaries*, **20**, 1–13.

Turner, D. D. (2004) The past vs. the tiny: Historical science and the ab⌐ ⌐ orking hypoth-
realism. *Studies in the History and Philosophy of Science*, **35**, 1–17. ⌐ guments for

Turner, D. (2005) Local underdetermination in science. *Philosophy of Scienc⌐*

Turner, R.K. and Daily, G.C. (2008) The ecosystem services framework and ⌐ 09–230.
conservation. *Environmental and Resource Economics*, **39**, 25–35. ⌐ apital

Umbgrove, J.H.F. (1946) Origin of continental shelves. *Annals of the Association c Petrol⌐*
Geologists, **30**, 249–253.

Umbgrove, J.H.F. (1947) *The pulse of the earth*. Martin Nihoff, The Hague.

Van Bemmelen, R.W. (1962) The scientific character of geology. *Journal of Geology*, **69**,
453–463.

Van Frassen, B.C. (1987) The semantic approach to scientific theories. In N.J. Nersessian
(ed.), *The process of science*. Martinus Nijhoff, Dordrecht, 105–124.

Van Inwagen, P. (1990) *Material beings*. Cornell University Press, Ithaca.

Van Nes, E.H. and Scheffer, M. (2004) Large species shifts triggered by small forces. *The
American Naturalist*, **164**, 255–266.

Vandermeer, J. (1993) Loose coupling of predator-prey cycles: Entrainment, chaos, and
intermittency in the classic MacArthur consumer-resource equations. *The American
Naturalist*, **141**, 687–716.

Vellinga, M. and Wood, R.A. (2002) Global climate impacts of a collapse of the Atlantic
thermohaline circulation. *Climatic Change*, **54**, 251–267.

Von Englehardt, W. and Zimmerman, J. (1988) *Theory of earth science*. Cambridge University
Press, Cambridge.

Walker, D. (2012) A Kuhnian defence of inference to the best explanation. *Studies in History
and Philosophy of Science*, **43**, 64–73.

Walker, W.E., Harremoes, P., Rotmans, J., Van der Sluijs, J.P., Van Asselt, M.B.A., Janssen, P.,
Krayer von Krauss, M.P. (2003) Defining uncertainty: A conceptual basis for uncertainty
management in model-based decision support. *Integrated Assessment*, **4**, 5–17.

Walling, D. E. (1983) The sediment delivery problem. *Journal of Hydrology*, **65**, 209–237.

Wang, Y.J., Cheng, H., Edwards, R.L., An, Z.S., Wu, J.Y., Shen, C.-C. and Dorale, J.A. (2001)
A high-resolution absolute-dated late Pleistocene monsoon record from Hulu Cave, China.
Science, **294**, 2236–2239.

Weisberg, M. (2004) Robustness analysis. *Philosophy of Science*, **73**, 730–742.

Welford, M.R. and Thorn, C.E. (1994) The equilibrium concept in geomorphology. *Annals of
the Association of American Geographers*, **84**, 666–696.

West, G.B., Brown, J.H. and Enquist, B.J. (1997) A general model for the structure and
allometry scaling laws in biology. *Science*, **276**, 122–126.

Whatmore, S.J. (2009) Mapping knowledge controversies: Science, democracy and the
redistribution of expertise. *Progress in Human Geography*, **33**, 587–598.

White, I., Mottershead, D.N. and Harrison, S.J. (1992) *Environmental systems*. Routledge,
London.

White, M. (1998). *Isaac Newton: The last sorcerer*. Fourth Estate, London.

228 • References   *and reality: An essay in cosmology.* Cambridge University

Whitehead, A. *al kinds. Philosophy,* **63**, 29–42.
Press, Cam *undamental processes in ecology: An Earth systems approach.*
Wilkerson, T.ss, Oxford.
Wilkinson, itesson, J.O.H. and Robinson, D.A. (2000) Measuring rates of surface
Oxford mapping microtopography: The use of micro-ersoion meters and laser
Williams weathering. *Zeitschrift fur Geomorphologie, Supplement Band,* **120**,
down
scan
51 Berlow, E.L., Dunne, J.A., Barabasi, A-L. and Martinez, N.D. (2002) Two
Willi separation in complex food webs. *Proceedings of the National Academy of*
99, 12913–12916.
.P., Lamb, A.L., Leng, M.J., Gonzalez, S. and Huddart, D. (2005) Variability
ganic δ^{13}C and C/N in the Mersey Estuary, U.K. and its implications for sea-level
construction studies. *Estuarine, Coastal and Shelf Science,* **64**, 685–698.
W.C. (1981) Robustness, reliability, and overdetermination. In M. Brewer and
B. ollins (eds), *Scientific inquiry and the social sciences.* Jossey-Bass, San Francisco,
14–163.
Wolman, M.G. and Gerson, R. (1978) Relative scales of time and effectiveness of climate in
watershed geomorphology. *Earth Surface Processes,* **3**, 189–208.
Wolman, M.G. and Miller, J.P. (1960) Magnitude and frequency of forces in geomorphic
processes. *Journal of Geology,* **68**, 54–74.
Wood, R.M. (1985) *Dark side of the Earth.* HarperCollins, London.
Woodward, J. (2000) Explanation and invariance in the special sciences. *British Journal for
the Philosophy of Science,* **51**, 197–254.
Woodward, J. (2003) *Making things happen.* Oxford University Press, New York.
Wooldridge, S.W. and Linton, D.L. (1933) The loam terrains of south east England and their
relation to its early history. *Antiquity,* **7**, 297–310.
Wright, C.W. (1958) Order and disorder in nature. *Proceedings of the Geologists Association,*
69, 77–82.
Yalin, M.S. (1992) *River mechanics.* Pergamon, Exeter.
Yokoyama, Y., Lambeck, K., De Dekker, P., Johnston, P. and Fifield, L.K. (2000)
Timing of the last glacial maximum from observed sea-level minima. *Nature,* **406**,
713–716.
Zong, Y. (2004) Mid-Holocene sea-level highstand along southeast coast of China. *Quaternary
International,* **177**, 55–67.
Zong, Y., Lloyd, J.M., Leng, M.J., Yim, W.W.-S. and Huang, G. (2006) Reconstruction of
Holocene monsoon history from the Pearl River Estuary, southern China, using diatoms
and carbon isotope ratios. *The Holocene,* **16**, 251–263.
Zwack, L.M., Paciorek, C.J., Spengler, J.D. and Levy, J.L. (2011) Characterizing local
traffic contributions to particulate air pollution in street canyons using mobile monitoring
techniques. *Atmospheric Environment,* **45**, 2507–2514.

Index

References in **bold** indicate tables and in *italics* indicate figures.